Water Audits and Leak Detection

AWWA MANUAL M36

Second Edition

FOUNDED 1881

American Water Works Association

MANUAL OF WATER SUPPLY PRACTICES—M36, Second Edition

Water Audits and Leak Detection

Project Manager and Editor: Melissa Christensen
Production Editor: Michael Malgrande

Library of Congress Cataloging-in-Publication Data

Water audits and leak detection -- 2nd ed.
x, 100 p. 17×25 cm. -- (AWWA manual ; M36)
Includes bibliographical references and index.
ISBN 1-58321-018-0
1. Water leakage. 2. Water--Waste--Measurement. 3. Water--Waste--Costs. I. American Water Works Association. II. Series.

TD491 .A49 no. M36 1999
[TD495]
628.1 s--dc21
[628.1'44'0287]

99-052127
CIP

Printed in the United States of America
American Water Works Association
6666 West Quincy Avenue
Denver, CO 80235

ISBN 1-58321-018-0

Printed on recycled paper

Contents

List of Figures

List of Tables

Foreword

Detecting and repairing leaks in water systems is an effective way to conserve water and save money. Leak detection programs conserve water that utilities pay to obtain, treat, and pressurize. When water leaks from a system before it reaches the consumer, water agencies lose revenue and incur unnecessary costs.

The California Department of Water Resources conducted a study with 60 local water utilities using cost-effective leak detection programs. These utilities controlled leaks at a cost averaging less than what a utility usually pays for water.

Water audit and leak detection programs have benefits in addition to the value of water control. Meter testing performed as part of the water audit frequently identifies customer meters that inaccurately record water use. Recalibrating these inaccurate meters results in increased revenues to the water agency. Early detection of leaks reduces the chance that leaks will cause major property damage. Better knowledge of the location of valves and mains allows utility personnel to react quickly when emergencies occur.

The first step in leak detection and repair is to learn whether or not it is cost-effective. This is done by conducting a water audit. Chapter 2 gives step-by-step instructions and sample forms for conducting a comprehensive, system-wide audit. Instructions for figuring increased revenues from large-meter repair are also included, as are suggestions for correcting problems in the water system.

Subsequent chapters describe leak detection and repair programs. These chapters tell how to decide (1) whether or not a leak detection survey is cost-effective and (2) how much time and what equipment are needed for the survey. Step-by-step instructions for conducting the survey are given. Forms to pinpoint leaks, estimate losses, record the costs of the survey and repair, and evaluate the cost-effectiveness of the program are also included.

The appendices provide blank forms to use when conducting an audit or detection/repair program. A description of leak detection equipment, instructions on how to calculate evaporative water loss, general information on meters, and a table providing formulas to convert US customary units to metric units are also included.

Acknowledgments

This manual is based on *Water Audit and Leak Detection Guidebook*, published as a joint venture by the California Department of Water Resources (DWR) Office of Water Conservation and AWWA's California/Nevada section.

Credit is due to the many California public water supply agencies that participated in the California water audit and leak detection program. These agencies field-tested the procedures contained in the manual.

Appreciation is extended to the many individuals who contributed to the original guidebook, including Edna Bailey, Ian Ballantine, Earl Bingham, Alice Brooks, Carl Buettner, Christopher Carr, John Curtis, Allan Dietemann, Max Ewington, Lynda Herren, Suzanne Kane, Alan Kentfield, William Kingston, Travis Latham, Gordon Laverty, Paul McMann, Donald Paff, Patti Palacios, Charlie Pike, Joseph Romano, Larry Sears, Susan Tatayon, David Todd, Philip Utic, Donald Vandekar, and Dean Wheadon.

Photographs in the manual were prepared by Community Consultants, Inc., the California Department of Water Resources, and the Louisville (Ky.) Water Company.

This manual was reviewed and approved by the AWWA Leak Detection Committee. Members of the committee, at the time of approval, were as follows:

D.A. Liston, Chair, Massachusetts Water Resources Authority, Chestnut Hill, Mass.
G.B. Cole, Vice-Chair, Pitometer Associates, Little Falls, N.J.
G.A. Kunkel Jr., Secretary, Philadelphia Water Department, Philadelphia, Pa.
F.S. Brainard Jr., F.S. Brainard and Company, Burlington, N.J.
T.G. Brown, Heath Consultants, Inc., Houston, Texas
R.G. Craft, Waterloo Water Works, Waterloo, Iowa
W.A. Finger, City of Greensboro, Greensboro, N.C.
J.G. Hock, Westchester Joint Water Works, Mamaroneck, N.Y.
T.D. Jakubowski, Rust Environment and Infrastructure, Milwaukee, Wis.
D.R. Leslie, North Park Public Water District, Machesney Park, Ill.
J.D. Liston, RadCom Technology, Malden, Mass.
W.E. Luta, Pennsylvania–American Water Company, Mechanicsburg, Pa.
T.J. McGee, Fluid Conservation Systems, Milford, Ohio
K.J. Nelson, National Water and Power Company, Orlando, Fla.
J. Thornton, Julian Thornton Associates, San Paulo, Brazil
D.A. Wheadon, International Water Consultants, Inc., Park City, Utah
A.R. Whittaker, South West Water Services, Exeter, England
G.N. Zelch, Louisville Water Company, Louisville, Ky.
D. Post, Staff Secretary, American Water Works Association, Denver, Colo.

Chapter 1

Introduction

A water audit, followed by a leak detection program, can help water utilities reduce water and revenue losses and make better use of water resources. This manual includes techniques for conducting water audits and leak detection surveys of piped, pressurized, potable water distribution systems. Major features of the manual are

- step-by-step procedures for conducting a comprehensive, system-wide water audit to assess the delivery efficiency of a distribution system
- a worksheet and sample forms for each step of the audit
- specific techniques to identify, measure, and verify all water sources, uses, and losses
- a procedure and form to calculate increased revenues as a result of testing and repair of large meters
- suggestions for correcting problems in a water system
- a tool to determine whether or not a leak detection project will be cost-effective, the length of time needed to perform a leak detection survey, and what equipment is needed
- step-by-step procedures for conducting a comprehensive leak detection project to locate nonvisible underground leaks
- forms to record information from surveying the distribution system, pinpointing leaks, estimating leak losses, and documenting costs of survey and repair
- a procedure and forms to evaluate the cost-effectiveness of the leak detection project

WHAT IS A WATER AUDIT?

A water audit identifies how much water is lost and what that loss costs the utility. Records and system control equipment (such as meters) are checked for accuracy. The overall goal of the audit is to help the utility select and implement programs to reduce distribution system losses.

To conduct a water audit, it is necessary to

- verify and update system maps
- test master and source meters
- verify and update source records for inflow, metered use (such as billing information), and unmetered use (including estimates for parks, community centers, government facilities, and firefighting)
- test residential, commercial, and industrial meters for accuracy
- inspect water measuring devices for proper sizing, installation, and operation
- field-check distribution controls and system operating procedures

Any of these tasks may be performed by utility staff, consultants, or both.

Water audits should be performed annually to help managers adjust priorities, monitor progress, identify new areas of system losses, and establish new maintenance goals. Updating a water audit is usually less expensive than the original audit. Utilities may also wish to consider a system of continuous monitoring to target leak detection efforts.

Cost of a Water Audit

The cost of a water audit is the sum of in-house work and field work. Total cost depends on the size of the service area to be audited; the completeness, currency, and accuracy of the utility's records, including meter-testing programs and records; and the extent to which utility staff or consultants are used to conduct the audit.

The most expensive task in a water audit is testing large meters. The audit's total field cost depends on the number of meters that must be tested to provide data for the entire system.

WHAT IS LEAK DETECTION?

Leak detection is a survey of the distribution system to identify leak sounds and pinpoint the exact locations of hidden underground leaks.

Cost of Leak Detection

The cost of leak detection includes the costs of equipment and personnel. A crew is needed to survey the system, and personnel are also needed to pinpoint leaks, estimate water losses, and provide documentation. The cost of a leak detection crew depends on whether the utility uses its own staff, a consultant, or both.

Leak detection costs range from $75 to $300 per mile of main. Costs for consultants range from $150 to $500 per mile of main. Variables include salaries for leak detection crews, the number of contact points to be surveyed, spacing of contact points to be surveyed, types of mains and services, accuracy of maps, and the type of leaks to be pinpointed. Sources for equipment can be found in appendix E.

Leak repair costs are not considered a direct cost of a leak detection program. Since leaks are continually discovered and repaired in the normal course of the utility's operations, leaks found by the leak detection program would eventually be repaired at some time in the future, sometimes under emergency conditions.

BENEFITS OF WATER AUDITS AND LEAK DETECTION

Water audits and leak detection programs can achieve substantial benefits, including the following:

- Reduced water losses. Water audits and leak detection are the necessary first steps in a leak-repair program. Repairing leaks saves money for the utility through reduced power costs to deliver water and reduced chemical costs to treat water.
- Financial improvement. A water audit and leak detection program can increase revenues from customers who have been undercharged, lower the total cost of wholesale supplies, and reduce treatment and pumping costs.
- Increased knowledge of the distribution system. During a water audit, distribution personnel become familiar with the distribution system, including the location of mains and valves. This familiarity helps the utility to respond quickly to emergencies, such as main breaks.
- More efficient use of existing supplies. Reducing water losses helps stretch existing supplies to meet increased needs. This could help defer the construction of new water facilities, such as a new well, reservoir, or treatment plant.
- Safeguarding public health and property. Improved maintenance of a water distribution system helps reduce the likelihood of property damage and safeguards public health and safety.
- Improved public relations. Consumers appreciate maintenance of the water system. Field teams doing the water audit and leak detection or repair and maintenance work provide visual assurance that the system is being maintained.
- Reduced legal liability. By protecting public property and health and providing detailed information about the distribution system, water audits and leak detection help protect the utility from expensive lawsuits.
- Reduced disruption to customers. More leaks are repaired on a planned basis rather than developing into major breaks that disrupt service.

Chapter 2

Conducting a Water Audit

This chapter outlines the basic steps involved in conducting a water audit (see Figure 2-1). Instructions for accomplishing each step are included. In some cases, more than one method is described for accomplishing a particular step, thus allowing for the various configurations of water systems. Each water supplier may choose the technique best suited to the system under study.

BEFORE STARTING

A number of decisions should be made before beginning a water audit. While discussion of all factors is beyond the scope of this book, three factors that influence the reliability of the study are discussed in the following paragraphs.

Water Audit Worksheet

A sample water audit worksheet is included in this chapter (Figure 2-2). This chapter describes each step of the audit, including what information is needed, how to get that information, and how to enter it on the worksheet. Examples accompany the instructions to show how the data are gathered and how they are entered on the worksheet. Figure 2-2 uses data from examples in this chapter; appendix A provides a blank water audit worksheet.

Instructions for entering information on the worksheet appear in blue.

Set a Study Period

A water audit is a study over time. Choose a time period that allows analysis and evaluation of total system water use. One month or even six months is too short a time to give an overall picture of water flow through the system (see Figure 2-3). A 12-month study period is recommended. Most utility records are kept by the calendar or fiscal year; either system makes 12 months of data available. However, a calendar

Before the Audit

A. Establish a worksheet
B. Set a study period
C. Choose an official unit of measure

The Audit

Task 1 — Measure the Supply

1. Identify and map sources
2. Measure the water from each source
3. Adjust figures for total supply

Task 2 — Measure Authorized Metered Use

1. Identify metered uses
2. Measure metered use
3. Adjust figures for metered use

Task 3 — Measure Authorized Unmetered Use

1. Identify unmetered uses
2. Estimate unmetered use

Task 4 — Measure Water Losses

1. Identify potential water losses
2. Estimate losses by type

Task 5 — Analyze Audit Results

1. Identify recoverable leakage
2. Figure the value of recoverable leakage
3. Figure the cost of recovering leakage
4. Calculate the cost of leak detection

After the Audit

A. Analyze the value of losses and corrective measures
B. Evaluate potential corrective measures
C. Update the audit
D. Update the master plan

Figure 2-1 Basic steps in conducting a water audit

WATER AUDIT WORKSHEET

For: County Water Co. Audit Study Period: Jan. 1 – Dec. 31 1998

Line	Item	Subtotal (Water Volume)	Total Cumulative (Water Volume)	Units*
Task 1—Measure the Supply				
1	Uncorrected total water supply to the distribution system (total of master meters)		3,672.36	mil gal
2A–C	Adjustments to total water supply			
2A	Source meter error (+ or –)	+82.90		mil gal
2B	Change in reservoir and tank storage (+ or –)	+0.83		mil gal
2C	Other contributions or losses (+ or –)	0		mil gal
3	Total adjustments to total water supply (add lines 2A, 2B, and 2C)		+ 83.73	mil gal
4	Adjusted total water supply to the distribution system (add lines 1 and 3)		3,756.09	mil gal
Task 2—Measure Authorized Metered Use				
5	Uncorrected total metered water use	3,258.0		mil gal
6	Adjustments due to meter reading lag time (+ or –)	+0.20		mil gal
7	Metered deliveries (add lines 5 and 6)		3,258.20	mil gal
8A–C	Total sales meter error and system-service meter errors (+ or –)			
8A	Residential meter error	134.33		mil gal
8B	Large meter error	29.97		mil gal
8C	Total (add lines 8A and 8B)		164.30	mil gal
9	Corrected total metered water deliveries (add lines 7 and 8C)		3,422.50	mil gal
10	Corrected total unmetered water (subtract line 9 from line 4)		333.59	mil gal
Task 3—Measure Authorized Unmetered Use				
11A	Firefighting and firefighting training	9.70		mil gal
11B	Main flushing	1.60		mil gal

*Units of measure must be consistent throughout the worksheet. The particular unit used (that is, gallons, millions of gallons, acre feet, cubic feet, cubic metres, or other unit) is left to the user.

Figure 2-2 Completed water audit worksheet for County Water Company (continued)

Line	Item	Water Volume: Subtotal	Water Volume: Total Cumulative	Units*
	Measure authorized unmetered use (continued)			
11C	Storm-drain flushing	0.50		mil gal
11D	Sewer cleaning	0.65		mil gal
11E	Street cleaning	1.75		mil gal
11F	Schools	0		mil gal
11G	Landscaping in large public areas:			
	Parks	40.00		mil gal
	Golf courses	120.00		mil gal
	Cemeteries	3.00		mil gal
	Playgrounds	5.40		mil gal
	Highway median strips	0.65		mil gal
	Other landscaping	0.50		mil gal
11H	Decorative water facilities	0 (metered)		mil gal
11I	Swimming pools	0 (metered)		mil gal
11J	Construction sites	0 (metered)		mil gal
11K	Water quality and other testing (pressure-testing pipe, water quality, etc.)	0 (metered)		mil gal
11L	Process water at treatment plants	0.07		mil gal
11M	Other unmetered uses	0		mil gal
12	Total authorized unmetered water (add lines 11A through 11M)		183.82	mil gal
13	Total water losses (subtract line 12 from line 10)		149.77	mil gal
Task 4—Measure Water Losses				
14A	Accounting procedure errors	11.63		mil gal
14B	Unauthorized connections	0.33		mil gal
14C	Malfunctioning distribution system controls	0		mil gal
14D	Reservoir seepage and leakage	+ 0.07		mil gal
14E	Evaporation	9.78		mil gal

*Units of measure must be consistent throughout the worksheet. The particular unit used (that is, gallons, millions of gallons, acre feet, cubic feet, cubic metres, or other unit) is left to the user.

Figure 2-2 Completed water audit worksheet for County Water Company (continued)

Line	Item	Water Volume Subtotal	Water Volume Total Cumulative	Units*
	Measure identified water losses (continued)			
14F	Reservoir overflow	0		mil gal
14G	Discovered leaks	21.88		mil gal
14H	Unauthorized use	0		mil gal
15	Total identified water losses (add lines 14A through 14H)		43.69	mil gal
Task 5—Analyze Audit Results				
16	Potential water system leakage (subtract line 15 from line 13)		106.08	mil gal
17	Recoverable leakage (multiply line 16 by 0.50)		53.04	mil gal

Line	Item	Dollars per Unit of Volume
18A–B	Cost savings	
18A	Cost of water supply	$ 690/mil gal
18B	Variable operation and maintenance costs	$ 153/mil gal
19	Total costs per unit of recoverable leakage (add lines 18A and 18B)	$ 843/mil gal

Line	Item	Dollars per Year
20	One-year benefit from recoverable leakage (multiply line 17 by line 19)	$ 44,712.72
21	Total benefits from recovered leakage (multiply line 20 by 2)	$ 89,425.44
22	Total costs of leak detection project	$ 24,215.00
23	Benefit-to-cost ratio (divide line 21 by line 22)	3.69

Prepared by:

Name John Smith

Title Distribution Manager Date 3/27/99

*Units of measure must be consistent throughout the worksheet. The particular unit used (that is, gallons, millions of gallons, acre feet, cubic feet, cubic metres, or other unit) is left to the user.

Figure 2-2 Completed water audit worksheet for County Water Company (continued)

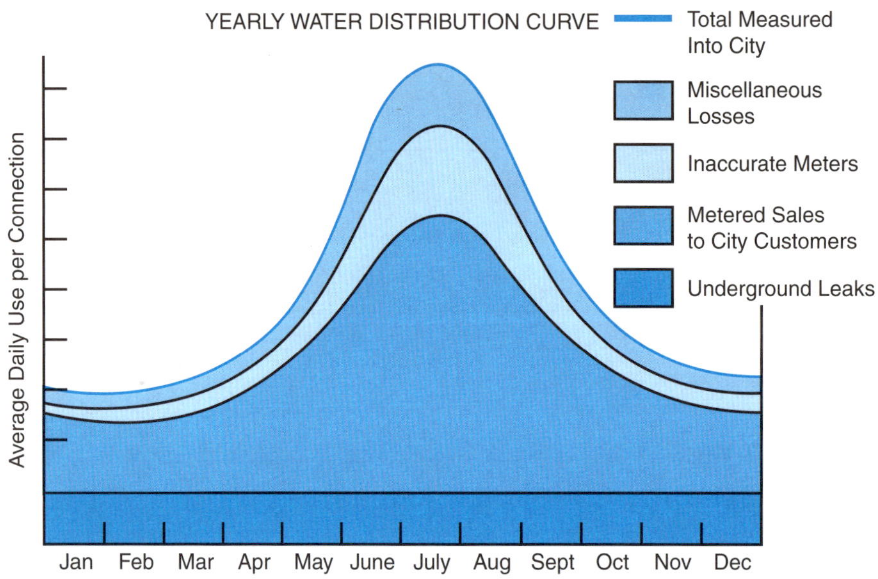

Figure 2-3 One calendar year is recommended for the audit study period because it includes seasonal variations

year is recommended to reduce the effects of lag time in meter reading, and it is long enough to include seasonal variations.

Choose a Unit of Measure

The same unit of measure should be used throughout the water audit. Most utilities in North America record total flow measurements in acre-feet, cubic feet, or gallons. Choose a unit of measure that suits the utility and use the same unit of measure throughout the audit.

In this book, the unit of measure is gallons. By way of comparison, 1 acre-ft equals 43,560 ft^3 or 325,851 gal.*

After establishing the water audit's official unit of measure, note the unit of measure used on each measuring device (that is, meter, weir, Parshall flume, and so on). Also note the conversion factor to be used when reading the device.

TASK 1—MEASURE THE SUPPLY

This task tells how much water enters the distribution system and where it comes from.

Step 1-1 Identify and Map Sources

1-1A Identify sources. Identify all water sources that supply the distribution system, including interconnections with other systems and intermittent sources or emergency supplies. Make a list of these sources.

*Refer to appendix F for US customary units to metric conversions.

Table 2-1 Source measuring devices for County Water Company

	Water Source		
	Source 1	Source 2	Source 3
Type of measuring device	Venturi	Propeller	Venturi
Identification number (may be serial number)	0000278-A	8759	OC-16
Frequency of reading	Daily	Weekly	Daily
Type of recording register	Dial	Dial	Builder type M
Units registers indicate	100,000 gal	Gallons	Cubic feet
Multiplier (if any)	1.0	1.0	100.0
Date of installation	1950	1968	1955
Size of conduit	24 in.	8 in.	11.5 in.
Frequency of testing	Annual	2 years	4 months
Date of latest calibration	4/1/91	8/21/91	1/15/92

1-1B Map the system. Find an existing map of the distribution system. The map should show the principal mains of the entire delivery system. A scale of 1 in. to 400 ft makes the map legible and easy to work with.

If no map of the distribution system exists, use aerial photos or a city or county map and draw the distribution mains on a transparent overlay.

If no suitable map can be found, a hand-drawn one can be used.

1-1C Plot the sources. First, choose a symbol to represent each type of water source, including aqueduct turnouts; wells; surface diversions, such as lakes, streams, or reservoirs; interconnections; and emergency sources. Then, draw the symbol on the map according to where the source is located.

Step 1-2 Measure the Water From Each Source

1-2A Identify measuring devices. Visit each source and note what type of measuring device is used (for example, meter, Parshall flume, weir, or stream gauge). (See appendix B for information on types of measuring devices and meters.) Note basic information about the measuring device, including the type, identification number, frequency of reading, type of recording register, unit of measure (and conversion factor, if necessary), multiplier, date of installation, size of conduit, frequency of testing, and date of last calibration. Using that information, construct a table similar to Table 2-1.

1-2B Record total water from each source. Record how much water was produced by each source for each month and for the entire audit period (see Table 2-2).

Most meters have some type of register, or totaling device. Registers may be round-reading or direct-reading. Round-reading registers have a series of small dials with pointers, registering cubic feet, or gallons, in tens, hundreds, thousands, and ten thousands. Direct-reading registers have one large sweep hand for testing and a direct-reading dial that shows total units of volume.

Calculate the total water produced from all water sources during the study period. Enter the amount on line 1 of the worksheet.

Table 2-2 Total water supply for County Water Company (uncorrected)*

Month	Source 1 Turnout 41 *mil gal*	Source 2 Well Field *mil gal*	Source 3 City Intertie *mil gal*	Total for Sources 1, 2, and 3 *mil gal*
January	0	130.34	104.27	234.61
February	0	195.51	65.17	260.68
March	0	260.68	0	260.68
April	130.34	130.34	0	260.68
May	265.57	97.76	0	363.33
June	299.78	0	81.46	381.24
July	303.04	0	84.72	387.76
August	325.85	0	89.61	415.46
September	293.27	32.59	32.59	358.45
October	130.34	32.59	97.76	260.69
November	130.34	0	130.34	260.69
December	130.34	0	97.76	228.10
Yearly Total	2,008.87	879.81	783.68	3,672.36†

*Study period is one calendar year.

†This is the total supply from all three sources for the year. Note that it is an *uncorrected* figure; if necessary, it will be adjusted.

Step 1-3 Adjust Figures for Total Supply

Figures for the total water supply, based on readings from source meters and measuring devices, are raw data. The raw data must be adjusted for a number of factors, including (1) meter inaccuracies, (2) changes in reservoir and storage levels, (3) nonmetered sources, and (4) losses that occur before water reaches the distribution system. These adjustments are made in the steps described below; they are recorded on lines 2A, 2B, and 2C on the worksheet.

1-3A Verify meter accuracy. Although most production sources are measured by meters, some are measured by other devices, such as Parshall flumes or weirs. Supply figures (like those used in Table 2-2) are based on readings of these measuring devices. Any error in any measuring device must be discovered and corrected; incorrect supply data invalidates the entire water audit.

To be sure meters are accurate, compare the results of meter tests to applicable AWWA standards. If a meter measures incorrectly and the error exceeds the standard for its category, repair and recalibrate the meter to function within standard limits.

If a meter has not been tested within the last 12 months, test the meter.

Possible causes of meter error. If source meters are inaccurate, inspect each one in the field. Normal wear is not the only cause of inaccurate meter readings. Check to be sure the meter is the right type and size for the application and that it is installed correctly. (See appendix B for types and uses of meters as well as sizing parameters and installation guidelines.) Check the size against manufacturers' recommended ranges. Be sure the meter is level; most meters are not designed for sloped or vertical operation. Inspect the meter to see if hard-water encrustation is interfering with its measurement.

Also check to verify that the proper registers were selected and installed correctly. Finally, be sure the register is read correctly. Have an employee other than the regular meter reader make a special reading of master meters, or have an employee accompany the meter reader to verify sample readings. Check to be sure the meter is read correctly, recorded correctly, and the correct conversion factor is used.

Check venturi meters. Check venturi meters for blockages in the throats of the meters. Test the primary device with a pitot rod. Testing the meter with a pitot rod shows whether or not the installation is adequate for nonturbulent flows. The meter's primary device should be tested at different flow ranges. If pressure deflection for appropriate flows is adjusted without checking the venturi itself, the meter may still record flows erroneously.

Testing meters. There are four ways meters may be tested. (Meter testing, surveillance, and calibration are discussed in appendix C.) Meter testing methods are listed here in order of effectiveness, with the most effective first.

1. Test the meters in place. Some pipes may need to be replaced to make this possible.
2. Compare meter readings with readings of a calibrated meter installed in series with the original meter.
3. Record meter readings for a given flow over a specified time period. Remove the meter and replace it with a calibrated meter. Record readings from the calibrated meter for the same flow over the same period; compare the readings.
4. Test the meter at a meter testing facility.

Meters can be tested with portable equipment. Pump efficiency flow testing can be used to check meters; it is sometimes provided free of charge by electric utilities. Some utilities use an averaging rod meter or anubar to test meters, but results may be off by as much as 10 percent. A standard single-point pitot rod must be used for accurate results.

Meter testing may be done by an outside agency. Consultants, meter manufacturers, and special testing laboratories offer testing services.

1-3B Adjust supply totals. Adjust the monthly and annual supply data from Table 2-2 for meter error. To do this, divide the uncorrected metered volume (UMV) by the measured accuracy of the meter (a percentage expressed as a decimal) and subtract the UMV as follows:

$$\frac{\text{uncorrected meter volume}}{\text{percent accuracy}} - \text{uncorrected meter volume}$$

$$= \text{corrected meter volume} \qquad \text{(Eq 2-1)}$$

Table 2-3 shows how to adjust the supply totals from Table 2-2 to yield the adjusted measurements.

Enter the total adjustments due to meter error on line 2A of the worksheet.

1-3C Adjust reservoir and tank storage. If source meters are located upstream of reservoirs and storage tanks, then stored water must be accounted for in the water audit. Generally, water flowing out of storage is replaced; as the "replacement" water flows from the source into storage, it is measured as supply into the system. If the reservoirs have more water at the end of the study period than at the beginning, then the increased storage is measured by the source meters but not delivered to consumers. Such increases in storage should be subtracted from the metered supply. Conversely, if there is a net reduction in storage, then the decreased amount of stored water should be added to the metered supply. Table 2-4 shows how to figure the change in storage volume.

Enter the changes in reservoir and tank storage on line 2B of the worksheet.

Table 2-3 Total water supply for County Water Company (adjusted for meter error)

Source	Yearly Total: Uncorrected Metered Volume (UMV)* *mil gal*	Meter Accuracy (MA) *percent*	Meter Error Calculation $\frac{UMV}{MA\dagger} - UMV$	Meter Error *mil gal*	Corrected Metered Volume‡ *mil gal*
1	2,008.87	95	(2,008.87/0.95) – 2,008.87	+105.73	2,114.60
2	879.81	100	(879.81/0.00) – 879.81	+00.0	879.81
3	783.68	103	(783.68/1.03) – 784	–22.83	760.85
Total adjustments due to meter error				+82.90	

*Based on Table 2-2.

†A percentage, written as a decimal (95 percent = 0.95).

‡The corrected meter volume for sources 1, 2, and 3 is 3,755.26 mil gal; note that this is 82.90 mil gal greater than the total supply given for these sources in Table 2-2. This is a way to double-check your arithmetic. The new total is not recorded on the worksheet—the "total adjustment due to meter error" is. This is only one of three adjustments that must be made to the raw data given in Table 2-2.

Table 2-4 Changes in reservoir storage for County Water Company

Reservoir	Start Volume *gal*	End Volume *gal*	Change in Volume *gal*
Apple Hill	32,350	36,270	+3,920
Cedar Ridge	278,100	240,600	–37,500
Monument Road	978,400	318,400	–660,000
Davis	187,300	55,300	–132,000
Total change in reservoir storage			–825,580 gal

Remember: *Decreases* in storage are *added* to the supply; storage *increases* are *subtracted* from the supply.

1-3D Other adjustments. Some water supplies may be subject to other types of contributions or losses. For example, there may be an additional source that enters the water system between the source meter and the finished water system. This could result from infiltration into an open channel. Likewise, losses may be introduced through an unlined or open channel. These additions or losses should be accounted for as "other contributions or losses" on the worksheet.

Enter other contributions or losses on line 2C of the worksheet.

Remember: Always use the same unit of measure.

Add lines 2A, 2B, and 2C; enter the sum on line 3 of the worksheet.

Add line 3 to line 1; enter the sum on line 4 of the worksheet.

TASK 2—MEASURE AUTHORIZED METERED USE

Authorized water is any water used for all uses approved by the utility. Most authorized use is metered, but some is not. Metered water, usually sold to consumers, includes industrial, commercial, residential, agricultural, governmental, and other uses. Unmetered water is used for irregular or mobile public purposes, such as street cleaning or firefighting. All unmetered uses should have meters installed if possible.

This task tells how much water goes to metered deliveries. The next task discusses unmetered deliveries.

Remember: To be accurate, the water audit must be consistent. Be sure to use the same 12-month study period and the same unit of measure for studying consumption as was used to study supply.

Step 2-1 Identify Metered Uses

2-1A Identify metered accounts. Identify all users who should have meters. Accounts can be identified by meter serial number, connection number, assessor's parcel number, street address, or account number. Assign each account to a meter-reading route.

Be sure to include all accounts for which data on metered use are available, even if the account is not billed. Remember, too, to take into account water provided to other agencies.

2-1B Describe meters. For all active accounts, list meters according to identification number and size of meter. Sort by type of use, including industrial, commercial, residential, agricultural, wholesale transfers, and other. This can help to identify accounts that represent larger volumes of sales and, therefore, greater potential earnings (see Table 2-5).

Check carefully to be sure all information is correct. Consider the possibility of accounting procedure errors, improper computer programming, incorrect meter reading, and unauthorized use. When data on metered use are unavailable or incorrect, water losses are possible. Task 4 gives methods for estimating potential water losses due to incorrect data on metered use.

Step 2-2 Measure Metered Use

2-2A Figure total (uncorrected) water use for each size of meter. Add the water consumption for all accounts and connections for each size of meter by month (or other billing period) and for the entire study period (see Table 2-6).

Calculate the total water consumed through all meters during the audit period. Enter the amount on line 5 of the worksheet.

Remember: Use the same unit of measure that was used for measuring supply in task 1.

2-2B Adjust for lag time in meter readings. Corrections must be made to metered use data when the source-meter reading dates and the customer-meter reading dates do not coincide with the beginning and ending dates of the audit study period.

Adjusting for one meter route. For example, a utility is studying one calendar year, January 1 through December 31. Source meters are read on the first day of each month and customers' meters are read on the 10th day of each month. The goal is to calculate the amount of water supplied and consumed for the calendar year.

Source meters. No correction is made for source meters, because their reading usually occurs on the days that the study period begins and ends. If the last reading (December 31) was a day late (January 1), then the water supplied for January 1 should be subtracted from the total water use read.

Customer meters. Because customer meter readings do not coincide neatly with the study period, a correction must be made. The best way to account for changes in the number of customers and in use patterns is to prorate water use for the first and last billing periods within the study period.

The first billing period has only 10 days that actually occur in the study period. Yet the billing information represents 31 days of use. If sales for that December 11

Table 2-5 Water consumption by meter size

Meter Size *in.*	Number of Meters	Percent of Total Meters	Percent of Metered Consumption
⅝	11,480	94.1	70.1
¾	10	0.08	0.1
1	338	4.4	2.8
1 ½	124	1.0	2.8
2	216	1.8	11.7
3	15	0.12	6.4
4	7	0.05	2.0
6	6	0.05	2.5
Total	12,196	100.00	100.0

Table 2-6 Total metered water use (uncorrected)

	Type of Use				
Month	Residential *mil gal*	Industrial *mil gal*	Commercial *mil gal*	Metered Agriculture *mil gal*	Totals for All Meters *mil gal*
January	146.6	35.8	8.1	0	190.5
February	162.9	35.8	8.1	0	206.8
March	162.9	35.8	8.1	0	206.8
April	179.2	39.1	8.1	24.4	250.8
May	211.8	42.4	8.1	57.0	319.3
June	228.1	48.9	8.1	74.9	360.0
July	260.3	48.9	8.1	57.0	374.3
August	266.5	48.9	8.1	74.9	398.4
September	228.1	45.6	8.1	65.2	347.0
October	162.9	35.8	8.1	0	206.8
November	162.9	35.8	8.1	0	206.8
December	146.6	35.8	8.1	0	190.5
Yearly Total	2,318.8	488.6	97.2	353.4	3,258.0

through January 10 period are 33.204 mil gal, the amount applicable to the study period is

$$33.204 \text{ mil gal} \times \frac{10 \text{ days}}{31 \text{ days}} = 10.711 \text{ mil gal} \qquad \text{(Eq 2-2)}$$

Thus, only 10.711 mil gal of the use read on January 10 applies to the study period.

At the end of the study period, there are 21 days not included in the billing data collected on December 10. Use for the last 21 days in December is obtained from the following month's billing. If sales for that month are 36.66 mil gal, the amount applicable to the study period is

$$36.66 \text{ mil gal} \times \frac{21 \text{ days}}{31 \text{ days}} = 24.83 \text{ mil gal} \qquad \text{(Eq 2-3)}$$

Thus, 24.83 mil gal is added to the use read on December 10.

Adjusting for many meter routes. The preceding discussion describes the basic method for correcting lag time in meter reading when all customers' meters are read on the same day. That seldom happens, however. Usually, meters are assigned to different routes and read on different days. Therefore, a meter lag correction should be used for each meter reading route, particularly if each customer's meter is read on the same date each month.

A meter lag correction can involve a number of steps. In our example, County Water Company has three meter routes, each with its own reading date. The study period is one calendar year, and the consumption is prorated for each meter route or book. Meters are read bimonthly: route A on the first of the month, route B on the 10th of the month, and route C on the 20th of the month (see Figure 2-4).

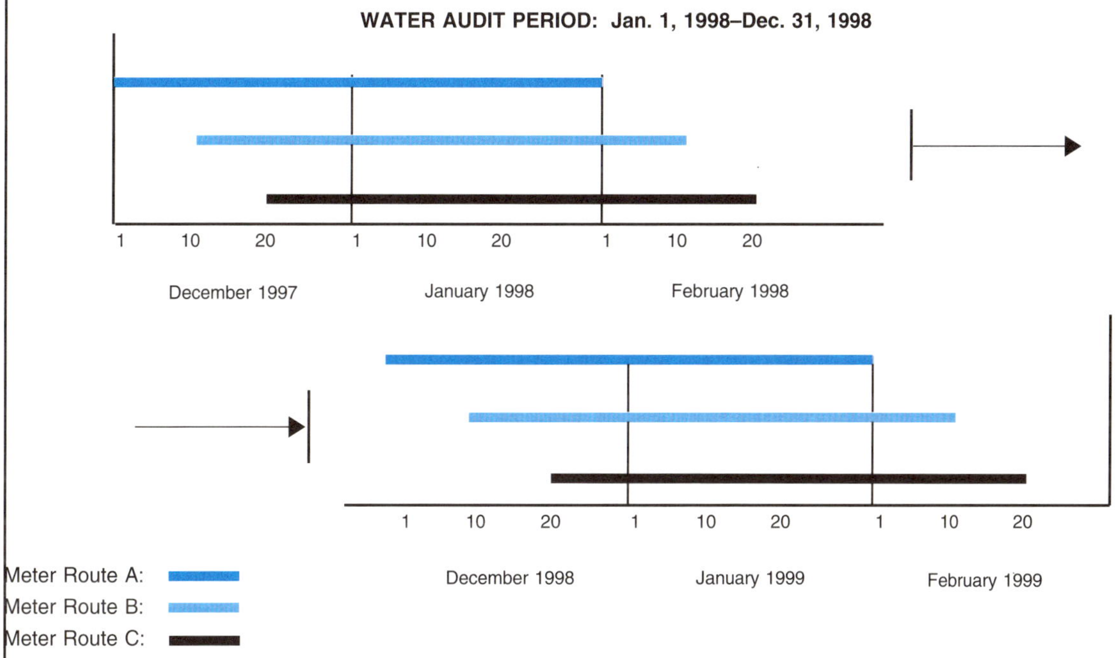

The December through January billing period is 62 days long.

Route	Date Read	Sales	Adjustment
A	2/1/98	4.0 mil gal	31/62 = 2.0 mil gal
B	2/10/98	3.3 mil gal	21/62 = 1.1 mil gal
C	2/20/98	3.6 mil gal	11/62 = 0.6 mil gal

Total adjustment to eliminate 1997 sales from study period . . . . –3.7 mil gal

This amount appeared on the February billing, but the water was used the previous December.

Route	Date Read	Sales	Adjustment
A	2/1/99	4.2 mil gal	31/62 = 2.1 mil gal
B	2/10/99	3.3 mil gal	21/62 = 1.1 mil gal
C	12/20/99	3.9 mil gal	11/62 = 0.7 mil gal

Total adjustment to add 12/98 sales to study period. . . . . +3.9 mil gal

This amount did not appear on the final bill for the year; it is prorated from the bill on which it appears.

Net adjustment . . . . +0.20 mil gal

Figure 2-4 Detailed meter lag correction

The uncorrected total metered use (from step 2-2A, Table 2-6) is based on bills issued during the study period. But, because of the bimonthly billing schedule, these bills would not include all water used during the year. Some water shown as used in the first billing period (issued in February) actually occurred in the preceding December. The last set of bills, issued in November and December, would not include water used in December. Two corrections need to be made. First, water used in the month preceding the study period must be subtracted from consumption figures. Second, water used in the final month of the study period must be added.

Figure 2-4 shows how to adjust sales figures for meter lag time. Many utilities combine accounting and billing procedures into a computerized format to make this procedure easier and quicker.

Prorate water sales figures to adjust for lag time in meter reading. Enter the net adjustment on line 6 of the worksheet.

Add all metered deliveries. Enter the sum on line 7 of the worksheet.

Step 2-3 Adjust Figures for Metered Use

Because there are so many customer meters, it is not practical to inspect and test every one each year. Instead, annual inspections and testing should include all meters greater than 2 in. in diameter, along with a random sample of smaller meters.

2-3A Check for proper installation. Review the utility's practices on meter selection, sizing, and installation to see whether or not present practices permit accurate operation. If they do not, revise the practices as necessary so that meters will operate correctly. (See appendices B and C for more information.)

Commercial and industrial meters produce a much larger share of revenue per account than do residential meters. Commercial and industrial accounts should be inspected for proper selection, sizing, and installation. In addition, inspect and test all large meters before they are used. Not all new meters are accurate.

2-3B Test residential meters. Test a random sample of residential meters—50 to 100 is a good number. Residential meters may be tested on a test bench or sent to the factory or a consultant for testing. (For more information, see appendix C of this book as well as Chapter 6 of *Water Meters—Selection, Installation, Testing, and Maintenance*, AWWA Manual M6.)

Meter replacement programs. Many utilities are involved in meter replacement programs. For those utilities, to calculate meter error for the entire system, the random sample of meters must include some of the meters being replaced. Test a representative sample of the new residential meters before putting them to use.

2-3C Calculate total sales meter error. Total sales meter error includes meter errors from all meter sizes, including residential, commercial, and industrial.

Calculate residential meter error. Residential meters are tested for low, medium, and high flows. The results, expressed as a percentage of accuracy, are used to calculate the total meter error at average flow rates. Tables 2-7 through 2-9 demonstrate how to use existing meter test data to calculate total residential meter error. The data in the table are based on Table 2-6.

Calculate large meter error. Tables 2-10 through 2-12 show how to use existing meter test data to calculate total large meter error. The mean registration data in Table 2-10 are used to calculate the meter error for large meters.

One of the benefits of a water audit is the potential increase in revenue resulting from testing and repairing large meters (performed as part of the audit). One can estimate the amount of revenue to be gained by repairing the meters; detailed instructions are given in appendix C.

Calculate total sales meter error. Total sales meter error includes meter errors from all meter sizes. In short, total sales meter error = residential meter error + large meter error. Using the data given in Tables 2-6 through 2-11, the total sales meter error for County Water Company is

$$134.33 \text{ mil gal} + 29.97 \text{ mil gal} = 164.30 \text{ mil gal} \quad \text{(Eq 2-4)}$$

Add lines 8A and 8B on the worksheet. Enter the sum on line 8C.
Add lines 7 and 8C on the worksheet. Enter the sum on line 9.
Subtract line 9 from line 4.

TASK 3—MEASURE AUTHORIZED UNMETERED USE

The volume of unmetered water must be carefully estimated to produce an accurate audit. This task includes descriptions of ways to measure water used for unmetered purposes. In selecting the best procedure for a given situation, consider the difficulty of gathering information; the degree of precision necessary; the availability of measuring equipment and skilled personnel; and the need for hiring consultants, buying more equipment, or training employees. When it is evident that a particular water use is very low, a rough estimate could replace a complete, detailed calculation.

It is recommended that all uses be metered, even if the customer is not billed for the use.

Procedures for Estimating Usage

Most unmetered water use can be estimated using either the batch or discharge procedure. Which procedure is best for a given use depends on how the water is applied. In some cases, the best way to estimate use is to adapt consumption figures from a similar, metered facility.

Batch procedure. When water is transported in a tank truck or container of some sort, use the batch procedure. Multiply the volume of the tank or other container by the number of times it is filled from the distribution system. This yields the volume of water delivered from the distribution system. Careful record keeping is necessary for accurate estimates.

Discharge procedure. When water is applied directly from a pipe, as in a sprinkler system, use the discharge procedure. Multiply the rate of water discharge by the total time it flows. This yields the volume of water delivered from the distribution system. The discharge rate may vary and the application period will vary in length and frequency. Again, careful record keeping is necessary for accurate estimates.

Comparison procedure. For some facilities and areas, such as schools, swimming pools, construction sites, or golf courses, consumption figures may be adapted from similar facilities, provided they are alike in size, hours and type of use, landscaping, and most other details. Any differences must be accounted for. For example, at a construction site, work habits are important. If the crew at a metered site turns off water between uses while the crew at an unmetered site lets the water run continuously, the borrowed consumption figures will have to be adjusted considerably.

Remember: No matter what procedure is used, be sure to use the same unit of measure used for other parts in the audit.

Table 2-7 Weighting factors for flow rates related to volume percentages for ⅝-in. × ¾-in. water meters*

Percent of Time	Range *gpm*		Average *gpm*	Percent Volume†
15	Low	0.50–1.0	0.75	2.0
70	Medium	1–10	5.00	63.8
15	High	10–15	12.50	34.2

*Based on information from Tao, Penchin, "Statistical Sampling Technique for Controlling the Accuracy of Small Meters," *Journal AWWA*, 6:296 (1982).

†Percent volume refers to the proportion of water consumed at the specified flow rate, as compared to the total volume consumed at all rates. In this example, only 2.0 percent of the total water consumed occurs at the low-flow range of approximately 0.5–1 gpm.

Instead of using the percentage of volumes shown here, you may compute your own percentage volume data. Using special dual-meter yokes and recording meters, you can determine the actual flow rates for your water meters.

Table 2-8 Meter testing data from a random sample of 50 meters for County Water Company

Test Flow Rates		Mean Registration *percent*
Low flow	(0.25 gpm)	88.8
Medium flow	(2.0 gpm)	95.0
High flow	(15.0 gpm)	94.0

Table 2-9 Calculation of residential water meter error

Percent Volume* (%V)	Total Sales Volume† (Vt) *mil gal*	Volume at Flow Rate (Vf) (%V × Vt) *mil gal*	Meter Registration (R)‡ *percent*	Meter Error (ME) ME = Vf/(0.01R) − Vf *mil gal*	Meter Error (ME) *mil gal*
2.0	2,318.8	46.38	88.8	[(46.38/0.888) − 46.38]	5.85
63.8	2,318.8	1,479.39	95.0	[(1,479.39/0.95) − 1,479.39]	77.86
34.2	2,318.8	793.03	94.0	[(793.03/0.94) − 793.03]	50.62
Total residential meter error (line 8A)					134.33

*From Table 2-7.

†Based on residential water sales data in Table 2-6.

‡From Table 2-8.

Table 2-10 Volume percentages for large meters for County Water Company*

Flow Rates	Percent of Volume Delivered
Low	10
Medium	65
High	25

*For this example, assume flow recordings were made for 24 h in July and February to indicate the percent of volume delivered by large meters at low-, medium-, and high-flow rates.

Table 2-11 Meter test data for large meters for County Water Company

Meter ID Number	Size *in.*	Meter Type	Date of Installation	Manufacturer	Test Date	Mean Registration at Various Flow Rates: (Designated as Percent of Registration) Low	Medium	High
XYZ001	3	Turbine	6/83	Sensus	10/91	89	85	100
X00ZAA	3	Turbine	6/83	Sensus	10/91	70	88	98
NB123	4	Displace	7/80	Sparling	10/91	95	99	102
NB456	6	Compound	9/77	Sparling	10/91	98	92.5	102
AA002	6	Propeller	5/66	Hersey	10/91	98	98	103
Sum of mean registrations						450	462.5	505
Mean registration for five meters tested						90	92.5	101

Table 2-12 Calculation of large water meter error

Percent Volume (%V)*	Total Sales Volume† (Vt) *mil gal*	Volume at Flow Rate (Vf) *mil gal*	Meter Registration (R)‡ *percent*	Meter Error (ME) ME = Vf/(0.01)R – Vf *mil gal*	Meter Error (ME) *mil gal*
10	488.6	48.86	90.0	[(48.86/0.90) – 48.86]	5.4
65	488.6	317.59	92.5	[(317.59/0.925) – 317.59]	25.75
25	488.6	122.15	101.0	[(122.15/1.01) – 122.15]	–1.21
Total meter error for large meters (line 8B)					29.97

*From Table 2-10.
†From Table 2-6.
‡From Table 2-11.

Step 3-1 Identify Unmetered Uses

Many water utilities provide some authorized, but unmetered, use of water. This is typically labeled “authorized, unaccounted-for water.” Most often, unmetered consumers are scattered throughout the service area at public buildings, open-space public areas, and special facilities designed to protect the public. Unmetered uses may include firefighting and training; flushing mains, storm drains, and sewers; street cleaning; schools; landscaping/irrigation in large public areas; wintertime bleeders; decorative water facilities; swimming pools; and construction sites. Water used for water quality and other testing, as well as process water at treatment plants, is also included in this category.

When conducting the audit, check to be sure these uses are unmetered. If they are metered, they should be handled as metered use, discussed in task 2.

Step 3-2 Estimate Authorized Unmetered Use

To estimate total unmetered use, break down the total into various uses.

3-2A Firefighting and training. This is defined as water drawn from hydrants, fire-sprinkler systems, and other water sources dependent on the piped water distribution system. It may be used for fire suppression, testing fire equipment, flushing sprinkler systems, or hazardous-materials reduction performed by public-safety crews. It may also be used to train firefighters, airport personnel, and other public-safety employees and volunteers. This category excludes water drawn from ponds, rivers, or other water supplies not connected to the piped water distribution system.

To estimate this use, check fire department records on training, flushing, and fire suppression. Many fire departments use more water for training than for fighting fires. Where flowmeters on standby fire systems show water use, the maintenance superintendent of the building may have fire or test records.

Some fire departments require a "run report" whenever a unit responds to a call. A survey of run reports for all fire calls made during the audit study period in the water service area should yield an excellent estimate of the amount of water used by the fire department. Remember to eliminate calls to locations where the water used came from water supplies not connected to the distribution system.

Estimates of other firefighting uses, such as sprinkler systems (including their testing), require calculation of the flow of the system and the duration of operation. For this calculation, the discharge procedure is used. To acquire the raw data needed for the calculation, survey and inspect meters at schools, stores, apartments, industrial sites, lumberyards, warehouses, and other similar locations. The more complete the survey, the more accurately the final estimate will reflect water used in testing, and in leaky or incorrectly connected sprinkler systems.

In our example, there are four fire companies in the service area. None of them make run reports. However, their logs show a total of 10 structural fires and a 5-day wildfire (for which water was airlifted from an open reservoir), plus 8 days (48 work hours) of training in which water was used. Estimates of water use are 6.5 mil gal for firefighting and 3.2 mil gal for training. Water used for fighting the wildfire is not included because it was not drawn from the distribution system.

Add fire-department use and other use to determine total use for firefighting and training. Enter the sum on line 11A of the worksheet.

3-2B Flushing mains. This water is vented from the distribution system, frequently to storm drains, to clean the system of contaminants and debris.

Many utilities with standard flushing procedures maintain logs that include the location of the main or blowoff and the length of time it flowed. Some utilities meter the amount of water released. (If this use is metered, the total should be included in the uncorrected total metered water use in task 2.)

Estimating water used for flushing mains requires a series of discharge estimates. For each location flushed, multiply the flow rate by the duration of the discharge. For instance, 50 gpm released for 30 min yields 1,500 gal. If the discharge rate is not constant, calculate the volume by figuring the area under a curve (see Figure 2-5).

In our example, mains were flushed through relief valves and hydrants on 18 occasions (54 h total), accounting for about 1.6 mil gal.

Enter the total volume used for flushing mains on line 11B of the worksheet.

3-2C Flushing storm sewers. This is water from the distribution system discharged through fire hydrants to flush and clean storm drains, culverts, or catch basins.

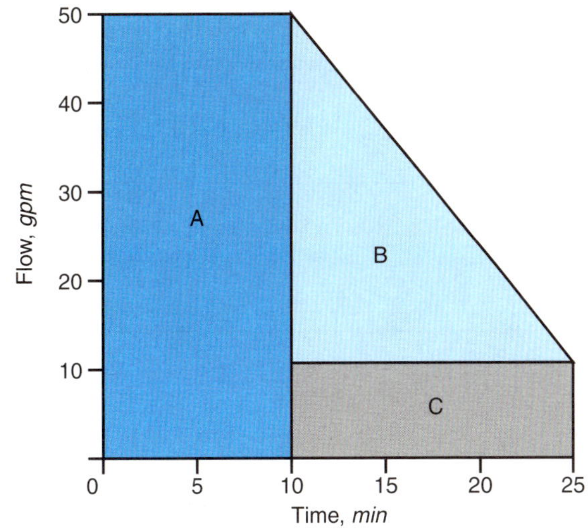

The discharge flow was constant for 10 min at 50 gpm, then uniformly reduced to 10 gpm over the next 15 min, and then was shut off.

Volume A	=	50 × 10 = 500 gal
Volume B	=	0.5 × (50 − 10) × (25 − 10) = 300 gal
Volume C	=	10 × 15 = 150 gal
Total volume	=	950 gal

Figure 2-5 Calculation of water volume from variable-rate discharge

To estimate water used for flushing storm drains, contact the department responsible for the work. Ask for logs of cleaning activity, including the number of trucks used, their capacities, how frequently they were filled, and portable meter readings. Use the discharge method if water is supplied directly from the piped distribution system. If water is transported by truck, use the batch method.

In our example, the county department of wastewater treatment estimated about 0.50 mil gal of water was used to clear congested storm drains.

Enter the total volume used for flushing storm drains on line 11C of the worksheet.

3-2D Flushing sanitary sewers. This is water from the distribution system discharged through fire hydrants to flush and clean sanitary sewers. This category also includes water used in sewage treatment plants for treatment processes and maintenance. This water should be metered and kept track of accordingly.

Use the procedure described in step 3-2C, for flushing water.

In our example, the county department of wastewater treatment estimated about 60 days' work using jet vactor and releases from hydrants for an estimated 0.65 mil gal of water.

Enter the total volume used for flushing sewers on line 11D of the worksheet.

3-2E Street cleaning. This water is used to clean roadways. It may be released directly from fire hydrants or sprayed from trucks, sweepers, or other equipment. It includes water used for cleaning park walkways, boat ramps, bus stops, parking areas, and bike paths.

Estimates are usually made with the batch method. Check with local street department to find out the number of trucks or other equipment used, their capacities, how many days they were used during the study period, and how many times a day they were filled. Also identify the types of equipment used to wash

Table 2-13 Estimate of water volumes used by tank trucks

Vehicle	Capacity *gal*	Number of Refills per Day	Number of Days Used per Year	Volume per Vehicle per Year *gal*
A	200	× 5	× 200	= 200,000
B	500	× 10	× 150	= 750,000
C	2,000	× 2	× 200	= 800,000
Total annual use.				1,750,000

recreational vehicle ramps and paths and the frequency of their use. Table 2-13 shows how to calculate total street-cleaning estimates.

Volumes used in direct cleaning from hydrants must be estimated using the discharge procedure.

Enter the total volume used for street cleaning on line 11E of the worksheet.

3-2F Schools. Water is used in schools for domestic sanitation, heating, and air conditioning. This designation may also apply to schoolyards and playgrounds that are supplied by school water services.

Estimate schools' use by comparing them with metered schools having similar landscaped areas and use characteristics, including the number of students and faculty, hours of use, and recreational facilities.

In our example, all schools are metered, so there is no unmetered use; a zero is entered on the worksheet.

Enter the total volume used for unmetered schools on line 11F of the worksheet.

3-2G Landscaping in large public areas. This water is used to irrigate parks, golf courses, cemeteries, playgrounds, highway median strips, and similar areas.

As with schools and other public areas, the easiest method of estimating this total water use may be to compare use with metered landscaped areas having similar use, watering schedules, size, and plant growth.

If records have been kept, it may be possible to calculate the actual amount of water applied to the landscaped areas. This would include runoff from misapplied watering. Ask the people who maintain the areas for information on the frequency and duration of watering. Remember that watering schedules vary at different times of the year.

For median strips and other landscaped areas watered by tank trucks, use the batch procedure for estimating volume.

For unmetered sprinkler systems, the discharge method can be used. Essential factors are (1) the discharge rate at each supply pipe to an irrigated area and (2) the total amount of time water is applied at each area. Obviously, time-controlled irrigation systems make the calculation easier. When figuring the amount of time water is applied, remember to use the total time the service is discharging, rather than the period for one lateral. Figure 2-6 demonstrates how to estimate the volume used for landscape irrigation. Figure 2-7 gives the landscape irrigation figures for the example, County Water Company.

Enter the total volume used for each category of unmetered landscaping on the lines under line 11G of the worksheet.

3-2H Decorative facilities. This water is used for cleaning and maintaining water quality in pools, fountains, and other decorative facilities.

EXAMPLE ESTIMATE OF PRIVATE LANDSCAPE WATERING

A single 2-in. service provides irrigation water to 4½-acre Sunnyslope Park at the rate of 160 gpm. Each of three laterals provides equal amounts of water and is controlled by a common timer.

Lateral A operates from 1:00 a.m. to 3:00 a.m. Lateral B operates from 3:00 a.m. to 5:00 a.m. Lateral C operates from 5:00 a.m. to 7:00 a.m. The system irrigates according to the following schedule:

May and September	Every third day
June	Every second day
July and August	Daily

How much water is applied from May through September?

Here's how to work out the answer:

The service supplies 160 gpm or 9,600 gallons per hour (160 × 60). It operates 6 hours each day the park is watered. During those 6 hours, 9,600 gallons per hour × 6 hours = 57,600 gallons of water applied.

The number of watering days must now be calculated:

Month	Days in Month	Frequency of Watering	Number of Days Watered
May	31	Every third day	11
June	30	Every second day	15
July	31	All days	31
August	31	All days	31
September	30	Every third day	11
Total			99 days

The total amount of water applied during the five-month period is

57,600 gpd × 99 days = 5,702,400 gal
= 762,353 ft^3
= 5.7 mil gal*

*The final answer must be given in the audit's official unit of measure.

Figure 2-6 Estimating landscape irrigation

The major causes of water loss from open-air, standing bodies of water are evaporation, water drained from a pool during maintenance, water used for cleaning, and leaks.

Evaporative loss. Appendix D explains how to calculate evaporative loss.

Pool drainage. To estimate water loss from pool drainage, use the following equation:

$$V \times F - V_w \quad \text{(Eq 2-5)}$$

Where:

V = volume of pool at the time it is drained (the volume of the pool when full minus the amount that is likely to have evaporated)
F = frequency of pool draining
V_w = volume of water loss due to drainage

Cleaning. To estimate the water lost in cleaning, ask maintenance workers about pool volumes and the frequency of cleaning and flushing.

EXAMPLE OF ESTIMATED PUBLIC LANDSCAPE IRRIGATION

Landscaping in large public areas is not metered. Three estimating methods were used.

1. Parks, Playgrounds, and Cemeteries.

For parks, playgrounds, and cemeteries comparisons were made with Mayfair City parks, playgrounds, and cemeteries, which are metered. Landscape was irrigated 2 mil gal/acre for parks and playgrounds and 1.5 mil gal/acre for cemeteries.

Type of Area	Total Area in Service District *acres*	Annual Water Rate *mil gal/acre*	Total Yearly Water Use *mil gal*
Park	20.0	2.0	40.0
Playground	2.7	2.0	5.4
Cemetery	2.0	1.5	3.0

2. Golf Courses.

Water use at the 62-acre municipal golf course was estimated by measuring the rate of flow and determining the length of watering periods.

Typical water application: 10 h/day
250 days/year
800 gpm

(10 h/day) (60 min/h) (250 days/year) (800 gpm) = 120 mil gal/year

3. Median Strips.

Highway median strips are watered by a large tanker truck carrying 500 gal, refilled 10 times a day, used 130 days a year.

$(500) \times (10) \times (130)$ = 0.65 mil gal/year

4. Miscellaneous.

A 25-acre memorial park is watered, but specific quantities are not recorded. Water use is estimated at 0.5 mil gal and included as "Other" on the worksheet.

Figure 2-7 County Water Company landscape irrigation figures

For an unmetered source, ask how much time maintenance work requires after the pool is drained. Ask if the hose or refill pipe is left running during that time. Determine flow rates for the appropriate outlet, refill pipe, or hose, and calculate the volume used. If the source is a hose bibb from a metered facility, no further calculation is needed.

Leaks. To estimate leakage, subtract the average amount that should be lost to evaporation from the normal water volume. The difference is leakage.

Add water lost to evaporation, drainage, cleaning, and leaks. Add losses by type of facilities (for example, parks or buildings) within the service area.

In our example, County Water Company has no unmetered decorative facilities. Enter the total volume used on line 11H of the worksheet.

3-2I Swimming pools. Here, water is used to maintain volume and water quality, including cleaning filters, decks, and walkways, and operation of sanitary and drinking water facilities associated with swimming pools. If concessionnaires tap into unmetered water intended for the pool, that use should be included, too.

Many pools are metered. In that case, their use is already counted as a metered use, under task 2. If they are not metered, estimate the volume of use from information provided by operations and maintenance staff. Useful estimates can be made by comparing water use with metered pools of similar size and use.

In our example, County Water Company has no unmetered swimming pools.

Enter the volume used for all *unmetered* pools on line 11I of the worksheet.

3-2J Construction sites. Water is delivered, principally through hydrants, to trucks for controlling road dust, site preparation, landscaping, temporary domestic use, and materials processing (for example, mixing concrete).

To estimate total use, use consumption data from metered construction sites for similar projects. Data may be obtained from regulatory water agencies. Compare the practice of shutting off supply at unmetered sites with the practices at metered sites and compensate for the difference.

It is recommended that all contractors be required to use a portable meter and report the readings.

In our example, County Water Company has no unmetered construction sites.

Enter the total volume used on line 11J of the worksheet.

3-2K Water quality and other testing. This water is used to test distribution system output to meet public health standards and to test meters and new mains.

Estimate water used by contacting operations staff to determine testing frequency as well as duration and volumes of water used. Amounts probably vary with each user.

In our example, County Water Company uses less than 500 gal/year for testing. The amount is so small it is disregarded.

Enter the total volume used on line 11K of the worksheet.

3-2L Process water at treatment plants. This is water lost—not recycled—after washing filters or draining sedimentation basins at source water treatment plants. If the meter is located after the treatment plant rather than before the plant, disregard this factor.

Estimate this water use by contacting plant operations staff and checking records.

In our example, operators for County Water Company do not keep specific records. It is estimated that a total of 0.07 mil gal is released to waste after filter backwash each year.

Enter the total volume on line 11L of the worksheet.

3-2M Other. An unmetered use may not fit any of the categories described here. In that case, determine the best means for estimating the total volume used.

In our example, there are no miscellaneous unmetered uses.

Enter the total volume on line 11M of the worksheet.

Add all unauthorized water uses (lines 11A through 11M). Enter the sum on line 12 of the worksheet.

TASK 4—MEASURE WATER LOSSES

This task shows how much water is lost from the distribution system, and how much of that is lost to leaks.

As we have seen, most water that passes through a distribution system is accounted for, that is, records show how much is used for what purpose. Accounted-for water could be defined as water that is either metered or used for an authorized, unmetered use. On the other hand, some water use is neither metered nor authorized. This water is considered lost from the system. The water does not produce revenue and is not available for beneficial uses. To determine how much water is lost by the system, subtract the volume of authorized, unmetered water from the corrected total unmetered water.

Subtract line 12 from line 10. Enter the difference on line 13 of the worksheet.

Step 4-1 Identify Potential Water Losses

Most water losses can be attributed to the following causes:

- accounting procedure errors
- unauthorized connections
- malfunctioning distribution-system controls
- reservoir seepage and leakage
- evaporation
- reservoir overflow
- unauthorized water use
- discovered leaks
- other leaks

The total volume of water lost to unknown leaks can be determined by first accounting for all other losses and subtracting them from the total water loss.

Step 4-2 Estimate Losses

This step identifies types of water losses and methods of estimating their volume.

4-2A Accounting procedure error. Water may seem to be lost from the system because of overlapping billing cycles (see discussion of meter lag time), misread meters, improper calculations, or computer-programming errors. These types of losses are paper mistakes that can be identified by careful, step-by-step review of record keeping and the computer process. For example, if the water meter registers in cubic feet and the water bill is in gallons, an incorrect conversion factor would introduce an error.

The entire billing and accounting procedure, from meter reading to printing billing statements, should be reviewed. In some agencies, the accounting and billing functions are scattered among several departments or organizational units. Such a fragmented arrangement can lead to miscommunication and error.

The task of reviewing billing and accounting can be simplified by checking a representative sample of accounts. A sampling of accounts from each meter-route book should be checked for accuracy. To do this, perform the following steps:

1. Determine the number of accounts to be checked from each book or route.
2. Choose a random sample of accounts to be checked.
3. Check meter readings. Have an employee other than the regular meter reader read meters for the identified accounts, or send another employee to accompany the meter reader to verify the readings. Both readings should be taken on the same day, if possible. The purpose is to determine whether or not the register is read and recorded properly and if the conversion factor is used properly. Compare the water-use volume for both meter readings. Calculate the billing manually.
4. Compare the total water use with total billed amounts. They should agree. If they don't, find out the reason for the discrepancy.
5. If meter-book readings show a substantial difference from billed amounts, review the normal billing process step by step, line by line.

In the example, four meter readers were accompanied on their rounds for a half-day each. This procedure sampled about 800 connections. In addition, the billing department's computer procedures were reviewed. The utility discovered that 0.5 percent of the 10,200 residential meters were inoperable or misread. The water loss was estimated at 0.228 mil gal per year for each inaccurate meter reading. The following equation shows the total water loss for the 1-year audit study period.

$$0.005 \times 10{,}200 \times 0.228 \text{ mil gal} = 11.628 \text{ mil gal} \qquad \text{(Eq 2-6)}$$

Enter the total loss due to accounting errors on line 14A of the worksheet.

4-2B Unauthorized connections. Identify active connections when unauthorized water use is not included in the accounting and billing process. Unauthorized use of water is usually accidental. For example, taps may be made to unmetered fire lines in a building (the installer may have been unaware that the lines are reserved for fire control); connections classified as inactive may be in use; or meters may not be read or the readings may not be entered into the accounting system. Occasionally, unauthorized water use may be deliberate. A customer may tap into the main or someone else's service to avoid paying for water. Good billing software should disclose improper practices, both deliberate and accidental.

The first step in measuring water lost to unauthorized connections is to determine which accounts are active and which are inactive. Connections listed as inactive should be confirmed. For some inactive accounts, the meter is removed or locked in the "off" position. If meters have not been deactivated, the meter reader should check the inactive connections periodically to see if they have been used. Telephone calls or visits to the premises may help meter readers keep abreast of changes in the use of these connections.

To identify unauthorized use, compare all active and inactive accounts with all locations that could receive water. One way to do this is to write the meter-identification numbers on the distribution system map and check that each parcel has a meter. A second method is to use an aerial photograph to identify places of water use and, where possible, correlate them with account numbers on a map. A third method is to check each meter-identification number against the assessor's parcel number. Any parcel without an account number should be investigated. (Keep in mind that some water uses may not have an account number or may not be metered.) A fourth method involves the measurement of flow into the suspected facility. A portable metering device is installed on the feed main upstream of the facility's meter. The flow is measured and recorded for a 24-h period. Results are compared to readings of the facility's meter. A discrepancy between the two amounts reveals that either the facility's meter is registering inaccurately or that water is being diverted through a bypass.

In our example, one small shop was using water illegally at an annual rate of 0.326 mil gal/year.

Enter the total volume for illegal connections on line 14B of the worksheet.

4-2C Malfunctioning distribution system controls. Water loss may result from improper application, malfunctioning, or improperly set system controls.

The basic steps for determining the volume of loss remain the same: determine (1) the rate of loss; (2) the length of time during which the loss occurred; and (3) the frequency of loss.

Valves. Valves are the controlling devices in a water distribution system. They are used for both isolation and control functions. Isolation valves are usually manually operated, while control valves function automatically. Many valves have indicators that show the position of the valve; valves without indicators can be retrofitted. Indicators make it easy to inspect valves to be sure they are positioned properly.

All control valves fail sooner or later. Each installation that uses automatic system controls should be inspected to determine whether or not the valve is working. Also check to see that the valve has been set properly and whether or not its size and design are suitable for its intended purpose.

Different kinds of valves must be checked for different problems.

Altitude control valves. Altitude-control valves can cause a tank or reservoir to spill if the valve is broken or set improperly. The valve is normally set to prevent the tank from overflowing.

Pressure-relief valves. If the pressure-relief valve is set too low for the system's range of pressures, then each time the pressure reaches the high range the valve will cause water to spill. Sometimes an unnecessary spill occurs because a pressure-regulating valve has been readjusted but the pressure-relief valve was not reset.

Pressure-reducing, pressure-sustaining, and pressure-maintaining valves. If any one of these valves is improperly set, it can cause an altitude-control valve, a pressure-relief valve, or a surge-control valve to spill water.

Surge-relief valves. If these valves are set too low, they can spill water as a blowoff to the atmosphere or into a tank or drain or back to the suction well of a pump.

Pump-discharge valves. When the pump-discharge valve fails, it acts like a check valve that is partially open. This may allow water to discharge from the distribution system down the well.

In this example, County Water Company uses an updated system schematic to methodically check all controls. All operated correctly.

Enter the total loss from malfunctioning distribution system controls on line 14C of the worksheet.

4-2D Reservoir seepage and leakage. This is the loss from linings, bottoms, or walls of storage tanks or ponds. Water loss is estimated by closing the inflow and outflow of the reservoir and noting the change in storage level over several days. From this information, the rate of seepage can be calculated. Because water may leak only at certain elevations, the seepage test should be performed several times at successively lower surface-water elevations. Clearwells (finished water storage) are also subject to seepage and should be inspected regularly.

In our example, each of County Water Company's closed reservoirs was individually valved off from the system for 24 h. One tank showed a change in elevation. The loss was calculated to be 192 gal/day, or 0.07 mil gal/year.

Enter the total loss from reservoir seepage and leakage on line 14D of the worksheet.

4-2E Evaporation. Most reservoirs that store treated water are covered and lined, which reduces evaporation significantly. Some clearwells and reservoirs are open to the atmosphere and subject to evaporation. Losses can be calculated by measuring the surface area and applying the proper evaporation data for the area (see appendix D).

In our example, Crystal Lake is an open reservoir with 5 acres of surface area. Estimated evaporation in excess of rainfall is 6 ft/year. Total annual loss is

$$6 \text{ ft} \times 5 \text{ acres} = 30 \text{ acre-ft, or } 9.776 \text{ mil gal/year} \qquad \text{(Eq 2-7)}$$

Enter the total volume lost to evaporation on line 14E of the worksheet.

4-2F Reservoir overflow. This occurs most often because the altitude control valve is faulty or missing. To calculate total loss, both the periods of overflow and the overflow discharge rate must be determined. (Overflow does not include water discharged to the distribution system.) If discharge is not directly measured (for

Table 2-14 Rate of leak losses

Diameter of Hole *in.*	Total Loss *gal/day*	Number of Days Leak Existed	Loss *gal*
1	243,000	30	7,290,000
1	243,000	30	7,290,000
0.25	15,200	180	2,736,000
0.25	15,200	300	4,560,000
Total loss from all leaks			21,876,000

example, by a stream gauge below the discharge point), reservoir overflow is calculated by subtracting reservoir outflow to the distribution system from the inflow to the reservoir. For open reservoirs, evaporation losses should be deducted from the reservoir inflow before attempting to calculate overflow.

In our example, no reservoirs overflowed.

Enter the total volume lost to reservoir overflow on line 14F of the worksheet.

4-2G Discovered leaks. Losses from leaks that are found and repaired can be measured to determine the rate of loss and the total volume lost during the life of the leak. Methods of estimating leak rates, with tables, are described in chapter 4.

In our example, four leaks were repaired. Two were large leaks believed to be short-lived (30 days), due to their disruptive nature. The other two were assumed to have been active the entire year until their discovery. Leak rates were not estimated in the field. The rates were estimated from the size of the hole in the pipe, using Greeley's formula as shown in Table 2-14. All mains involved operated at 50 psi.

Enter the total loss from discovered leaks on line 14G of the worksheet.

4-2H Unauthorized water use. Most unauthorized water use occurs when individuals vandalize fire hydrants or open the hydrants to fill water trucks. Comparing construction permits with temporary-use billings shows where large amounts of water are being taken without metering and billing.

In this example, there was no unauthorized water use.

Enter the volume lost to unauthorized water use on line 14H of the worksheet.

Add all the identified water losses (lines 14A through 14H). Enter the sum on line 15 of the worksheet.

TASK 5—ANALYZE AUDIT RESULTS

Audit results may indicate loss problems resulting from faulty metering, unauthorized taps, leaking reservoirs, or leaking mains and services. To determine which corrective efforts are cost-effective, first estimate the value of the recoverable leakage and the cost of recovering it. If the value of the recoverable water exceeds the cost of recovering it, then carry out a metering and unauthorized tap program and a leak detection and repair program.

This section shows how to figure the benefits and costs of a leak detection survey.

Step 5-1 Identify Recoverable Leakage

Not all leakage is recoverable. To determine what portion may be recovered, first find out how much leakage there probably is, then figure what percentage of that can be recovered.

5-1A Potential leakage. Generally, more water is lost than can be accounted for. This missing water is called potential leakage. It is easily calculated by subtracting identified losses from total losses.

Subtract line 15 from line 13. Enter the difference on line 16 of the worksheet.

5-1B Recoverable leakage. Not all leaks can be detected and repaired. Experience indicates that about 25 to 75 percent of all potential losses can be recovered.* To calculate the recoverable leakage, multiply the potential leakage by 0.50.

Multiply line 16 by 0.50. Enter the total on line 17 of the worksheet.

Step 5-2 Figure the Value of Recoverable Leakage

Saving water saves money. What cost savings would be achieved if the leakage was prevented? There are two types of cost savings: (1) the cost of purchasing the water and (2) variable operations and maintenance costs associated with storing, treating, and delivering the water. Both of these costs vary with the amount of water going into the distribution system. Both exclude fixed costs. The cost savings equal the value of recoverable leakage.

5-2A Purchase cost. This is what it costs the utility to buy water from another water supplier. Recovering leakage reduces the amount of water purchased. Usually, the most effective cost reduction results from reducing the amount of water purchased or produced from the most expensive source of supply.

In our example, the purchase price of water from the CC Aqueduct is $690/mil gal.

Enter the cost-per-unit of water from the utility's most expensive supply on line 18A of the worksheet.

5-2B Operations and maintenance. Operations and maintenance costs are paid to treat and pressurize water in the system. If the utility pumps water from its own wells, then recovering leakage will reduce the amount of energy needed for pumping.

NOTE: Only operation and maintenance costs that vary with the amount of water delivered are to be included here. Fixed costs that do not vary with the amount of water delivered should not be included.

In our example, $137.7/mil gal is spent to purchase power, while $15.3/mil gal is spent on water treatment. Variable operations and maintenance costs total $153/mil gal.

Enter the unit costs of variable operations and maintenance on line 18B of the worksheet.

5-2C Total cost per unit. The total cost per unit is the sum of costs for purchase and costs for oprations and maintenance.

Add lines 18A and 18B. Enter the sum on line 19.

Step 5-3 Figure the Cost to Recover Leakage

This step is accomplished in two parts.

5-3A One-year benefit. To determine the benefit of leak repair over one year, multiply the recoverable leakage by the unit cost of recoverable leakage.

Multiply line 17 by line 19. Enter the product on line 20 of the worksheet.

---

*According to the California Department of Water Resources' *Statewide Leak Detection Grant Program Final Report* (1988). For 47 agencies, the average recoverable leakage was 48 percent of the unidentified water losses indicated by audits.

5-3B Two-year benefit. The average lifetime of a leak before it requires repair is estimated to be two years, depending largely on pipe and soil conditions and the extent of a leak detection program. The total benefit, then, accrues over those two years. To compute total benefits from recoverable leakage, multiply the 1-year benefit by 2.

Multiply line 20 by 2. Enter the product on line 21 of the worksheet.

In some situations, it may be appropriate to include additional benefits from increased leak detection and repair. For example, a more active detection program may allow investments in new resources or delayed construction of a treatment plant, resulting in substantial financial savings from deferred capital spending.

Step 5-4 Calculate the Cost of Leak Detection

The cost of conducting a leak detection survey can be estimated by preparing a leak detection and repair plan. Chapter 4 provides suggestions for how to prepare the plan. Prepare the plan now to get an idea of the cost for leak detection.

Enter the cost for leak detection on line 22 of the worksheet.

NOTE: The cost of leak repair is not included. Since leaks are continually discovered and repaired in the normal course of utility operations, leaks found in the leak detection program would be repaired eventually. If leaks are repaired as part of the leak detection program, the utility avoids the expense of repairing them as they are discovered accidentally. Savings on future repair costs are often overlooked when estimates of savings from leak detection are made, but they can be nearly as great as the cost of repairing leaks as part of the program. The real cost of repairing leaks in a program is generally very small.

For example, when the average life of the leak is 24 months and the real interest rate is 3 percent, benefits from avoided future repair costs amount to 94 percent of the cost of repairing leaks at the time of the leak detection program. In other words, the real cost is only 6 percent of the cost of repairing the leaks found in the program. To simplify the calculation, the cost of repairs has been assumed to be 0.

To determine the benefit–cost ratio, divide total benefits from recoverable leakage by the total costs of the leak detection program.

Divide line 21 by line 22. Enter the total on line 23 of the worksheet.

If the benefit–cost ratio is greater than 1.0, then the benefits of leak detection are greater than the costs, and the program should be implemented.

AFTER THE AUDIT

Outlined here are steps management can take to correct the unproductive elements in the water supply system.

Analyze the Value of Losses and Corrective Measures

Evaluations of corrective measures should be based on cost, feasibility, and savings. The water audit tells the utility where the greatest losses occur. This information allows the utility to set priorities. In setting priorities, managers need to incorporate local constraints.

The choice of corrective actions is influenced by a number of factors, including the following:

- where losses occur
- how much water is lost in each problem area
- what action is needed to reduce water loss

- the cost of reducing water loss
- savings that will result from reducing water loss (this is based on the benefit–cost ratio)
- timetable for implementing the measures

Evaluate Potential Corrective Measures

Corrective measures include performing a leak detection survey and leak repair program, replacing mains that have a history of serious leaks, exercising valves annually, and implementing corrosion-control procedures.

Based on our example, County Water Company, the following recommendations might be made: conduct a leak detection program; test customer meters when received from the manufacturer; check accounts for stopped meters; recommend that public parks, playgrounds, cemeteries, and golf courses be metered; have the leak repair crew measure leak discharge when leaks are uncovered; and perform water audits annually.

Update the Audit

Annual updates provide information to help managers adjust priorities and monitor progress made on system maintenance. Equally important, the update can identify new areas of system losses; this helps managers establish new annual maintenance goals. Updating a water audit is usually less expensive than the original audit.

In some cases, it may be worthwhile to install permanent flowmeters in distribution zones. Continuous monitoring allows detection teams to be sent to areas of highest leakage and provides clear results of detection and repair.

Update the Master Plan

The utility's master plan can be used to set priorities and schedule corrective actions to maintain the distribution system. Managing a distribution system requires current information on the system's delivery capacity, maintenance, and water quality. An updated master plan, supplemented by a water audit and leak detection program, supplies that current information by providing the following:

- identification of problem areas and areas of potential water savings
- analysis of water and cost savings achieved by corrective action
- feasibility analysis of corrective actions based on cost and organizational constraints
- analysis of improved water system efficiencies resulting from past and proposed corrective actions
- analysis of greater system efficiencies versus expansion of resources, treatment plant, or distribution system
- projected water needs
- an implementation schedule for corrective actions
- updated maps showing the system's physical relationships and characteristics

Chapter 3

Leak Detection Overview

TYPES OF LEAKS AND THEIR COSTS

There are two basic types of leaks: visible and nonvisible. Visible leaks can be seen emerging from the ground or pavement (see Figure 3-1). The source of the leak may be a considerable distance away from the area where it is observed. Many visible leaks are reported by water customers.

Nonvisible leaks may percolate into the surrounding ground or may enter storm drains, sewers, stream channels, or old abandoned pipes (see Figure 3-2). In California, it is estimated that nonvisible leaks account for as much as 81,462.750 mil gal per year.

Costs of Underground Leaks

Water takes the path of least resistance; it may never surface. Most leaks start small and grow larger over time. Most nonvisible underground leaks are estimated to have an average life of two years, depending mostly on soil and pipe conditions and the extent of a leak detection program. As water leaks, it creates an underground cavity. This increases the potential for damage to overlying property. There are many examples of leaks that never surface. The following paragraphs illustrate a few of them.

City of Beverly Hills. An 8-in. diameter steel main with a pressure of 80 psi had a leak from a ½-in. hole in the top of the main. No evidence of leakage was visible. Flow from the leak was calculated to be 53 gpm. The water lost from this leak for the estimated 2-year duration was 55.720 mil gal. The value of the water lost was calculated to be $43,000.

Placer County Water Agency. The leak survey crew detected leak sounds on a fire hydrant served from a 4-in. cast-iron main. Using the ground microphone, the leak location was pinpointed and found inside an 18-in. concrete culvert. The main crossed through the inside of the culvert pipe. The culvert pipe drained storm runoff

Figure 3-1 Visible leaks, similar to those at this leaky fire hydrant, are easily found

Figure 3-2 Nonvisible leaks usually occur underground

from a hillside, crossed a road, then terminated in a ravine. Debris carried through the culvert over the years had damaged the 4-in. main and caused an 18-gpm leak that drained into the culvert. Over the 2-year period, 18.899 mil gal of water valued at $2,900 was lost through the leak.

City of Petaluma. The leak detection crew detected leak noise at several services. Using the ground microphone, the crew pinpointed the leak on an abandoned galvanized service line that had not been shut off. The water lost from the 5-gpm leak for the 2-year duration was 5.214 mil gal. The value of the water lost was $4,250.

Walnut Valley Water District. The leak detection crew detected a leak sound on a hydrant and service meter. The leak was pinpointed by using a correlator and found in a 1-in. polyvinyl chloride (PVC) service lateral near the connection to a 10-in. asbestos–cement (AC) main line. Pressure in the line was measured at 120 psi. The leak was a split in the PVC service lateral and was measured using a bucket and stopwatch at 12 gpm. The water lost from the leak for the 2-year duration was 12.708 mil gal. The value of the water lost was $9,600.

City of Santa Clara. The leak crew detected a leak sound on a 6-in. ductile, iron pipe. The leak, pinpointed on the main, was a circumferential break caused by settlement. Water from the main was leaking into an adjacent sewer at 98 gpm. Water lost from the leak for the 2-year duration was 103.295 mil gal. The value of the water lost was $47,000.

Value of Recoverable Leakage

Table 3-1 shows the value of recoverable leakage based on an average estimated leak lifetime of 2 years and varying leak rates and water costs.

OVERVIEW OF LEAK DETECTION METHODS

Listening for the leak itself is the direct method of detection. This is performed by a distribution system worker wearing listening equipment. There are other methods of leak detection, including zone flow measurement and the water audit. Rather than pinpoint leaks, these methods indicate whether or not water is leaking and give a general idea of where the leak might be. Another means of leak detection is not really a method, but chance—the accidental discovery of leaks during the normal course of distribution system maintenance.

Water Audit

A water audit is an efficient way to determine the total volume and value of water leaking from a distribution system. The final part of the water audit is preparing a leak detection and repair plan, which outlines the equipment, type of crew, method of surveying and pinpointing leaks, and the costs involved. Chapter 4 discusses the leak detection and repair plan in detail.

Audible Leak Detection

Audible leak detection uses electronic listening equipment to detect the sounds of leakage. Pressurized water forced out through a leak loses energy to the pipe wall and to the surrounding soil area. This energy creates audible sound waves that can be sensed and amplified by electronic transducers or, in some cases, by simple mechanical devices. The sound waves are evaluated to determine the exact location of the leak. During the leak detection survey, a trained operator conducts an initial listening survey of the entire distribution system and records all suspect sounds.

Table 3-1 Value of recoverable leakage

Leak Rate *gpm*	Cost of Water *$/thous gal*	Value of Leaking Water *$/2 years*
¼	0.30	79
¼	0.75	197
¼	1.50	394
¼	2.30	604
1	0.30	315
1	0.75	788
1	1.50	1,577
1	2.30	2,418
5	0.30	1,577
5	0.75	3,942
5	1.50	7,884
5	2.30	12,089
25	0.30	7,884
25	0.75	19,710
25	1.50	39,420
25	2.30	60,444
100	0.30	31,536
100	0.75	78,840
100	1.50	157,680
100	2.30	241,776
500	0.30	157,740
500	0.75	394,350
500	1.50	788,700
500	2.30	1,209,340

Later, areas with sounds are rechecked. If sounds can still be heard, leaks are pinpointed.

Types of leak sounds. There are three typical leak sounds. The first is in the 500 to 800 hertz (Hz) range. It usually originates as an orifice–pipe vibration phenomenon, and it is transmitted along the pipe wall and in the water—in some instances, a considerable distance from the actual leak. Identifying this sound by systematically testing valves, hydrants, and curb valves frequently locates potential leaks.

The second and third types of leak sounds are in the 20 to 250 Hz range. The second sound is caused by the impact of water on soil in the area of the leak. The third sound resembles the sound of a fountain. It is caused by water circulation, usually in a cavity in the soil near the leak. Unlike the vibration on the pipe wall, the travel distance of these two sounds is limited to the immediate area of the leak. Because of their limited range, these two sounds are essential to pinpointing the leak.

Factors affecting leak sounds. A number of factors influence leak sounds, including the following:

- Pressure. It is usually necessary to have 15 psi or more water pressure for sonic leak detection.
- Pipe material and pipe size. Sonic techniques can be used on pipe and fittings of any material. Because metallic pipe is a much better sound conductor than

nonmetallic pipe, a closer test interval is required when searching for leaks on nonmetallic pipe.

- Soil type. The type of soil greatly influences the amount of sound transmitted to the surface. Empirical observation indicates that sand is normally a good conductor of sound while clay is a poor conductor.
- Surface type. The type of surface on which the sounding instrument is placed also influences how the sound travels. Sod tends to insulate and muffle sounds, while asphalt and concrete are good resonators providing a uniform sounding surface.

Zone Flow Measurements

This method can be used as an extension of the water audit or, in some cases, as a leak detection method. Its purpose is to determine whether or not a sector or zone of a water system is experiencing major leakage. To effectively conduct a zone flow measurement, a utility must maintain good maps, have valves located at zone-control points, and provide a tap in the main for a pitot rod attached to a recording device.

To measure a zone, close valves to isolate sections of the water system and allow flow through a single line. Record flow for 24 h. Compare the average hourly rate to the minimum nighttime hourly rate. The resulting ratio may indicate leakage if the ratio is over 0.40 and there are no large industrial or commercial nighttime water users in the area. For example, total 24-h flow is 48,000 gal divided by 24 equals 2,000 gal average hourly flow. Minimum nighttime hourly flow of 1,000 gal divided by 2,000 results in a ratio of 0.50. Such zones, or districts as they are usually called, warrant additional investigation and correction.

Zone measurements are most effective in large residential or rural areas, and less effective in commercial or industrial areas. Winter months are best for zone measurements because usage is less affected by outside water use (that is, lawn watering and irrigation). However, measurements on nights with extreme freezing temperatures should be avoided, because customers may leave water running to keep water lines from freezing. Twenty-four-hour recordings are recommended, and daytime flows are compared with nighttime flows. Water utility staff should be alert to nighttime reservoir filling or unusual water usage, such as night irrigation, or a commercial use, such as laundry operation. Allowances are made for minimum use. Flows greater than these minimum flow rates indicate leakage.

Flow recorders can provide an accurate picture of system leakage by subtracting the known nighttime use from the minimum nighttime flow. Using a scaling factor based on the number of properties or the length of mains in a zone helps to compare leakage levels between different zones.

Another form of zone measurement is the storage-tank method. This type of zone measurement can be used in pressure zones where there are only a few customers and the entire zone is supplied by storage tanks. Each storage tank is read for volume, and all the customer meters in the zone are read. During that time period, storage tanks are not replenished. After a period of time, such as a week, the procedure is repeated. The potential leakage is the difference between the water supplied to the system from the storage tanks and the sum of the meter readings. (For this method to work, customer meters must be accurate. They should be tested before measurement begins.)

Normal Course of Operation

Utilities discover leaks accidentally in the normal course of operations and maintenance, for example, during a valve-exercising program. Meter readers also have the opportunity to check for visible meter-box leaks when reading meters.

LEAKS YOU CAN EXPECT TO FIND

Leaks usually can be divided by type into six categories based on where they occur. Leaks may be located on the main, the service line, a residential meter box, residential service, or in valves. Miscellaneous leaks also occur elsewhere in the system. Causes of leaks include improper installation, settlement, overloading, and other factors.

Main Leaks

Main leaks range from a low of 1 gpm to over 1,000 gpm. Leaks due to corrosion usually start out small but can grow to large leaks. Splits can occur due to excessive pressure, improper installation, settlement, and overloading. Joint leaks can occur because of corrosion, improper installation, improper materials, or overloading.

Service-Line Leaks

Service-line leaks range from a low of 0.5 gpm to over 15 gpm. Service-line leaks are caused by the same factors as main leaks.

Residential Meter-Box Leaks

Leaks near the meter box range from less than 1 gpm to 10 gpm. Leaks may be caused by loose spud nuts on the meter, loose packing nuts, damaged or broken angle stops, broken or damaged couplings, broken meters, or damaged or broken meter yokes.

Residential Leaks

Leaks on the residential customer side of the system range from less than 1 gpm to 15 gpm. These leaks may be caused by holes or breaks in customer service lines, inefficient hose-bibb or shutoff valves, holes or breaks in interior plumbing lines, or leakage inside plumbing fixtures (toilet-fixture leaks are common).

Valve Leaks

Leaks in distribution system valves range from less than 1 gpm to 500 gpm. Loose packing and broken valves are common causes of this kind of leak. Valve leaks may start in system controls, such as pressure-reducing valves, pressure-sustaining valves, pressure-relief valves, altitude-control valves, blowoffs, air-release valves, and others.

Miscellaneous Leaks

Excessive pressure, settlement, overloading, improper installation, improper materials, and improper operation may also cause breaks in water mains.

Chapter 4

Conducting a Leak Detection Survey

Leak detection is a process of elimination and discovery. The goal is to eliminate the contact points where leak sounds are not heard and discover the contact points where leaks can be heard. A contact point is any suitable connection to the water main that transmits sound vibrations. This can be a fire hydrant, service stop, valve, or probing rod.

PREPARING FOR THE LEAK DETECTION SURVEY

Before conducting a leak detection survey, review the specifics of the distribution system, including the following:

- results of the water audit: How much water is lost from the system?
- mains and services: types, ages, diameters, joints, installation methods, inspections, leak histories, and operating pressures
- meters and meter-box assemblies: types, brands, and sizes of meters; ages; types of installations; meter shutoffs; couplings; and meter reading frequency
- valves: locations, types, left- or right-handed, number of turns to exercise, and how often exercised
- hydrants: types, sizes, locations, flushing frequencies, and unmetered usage
- pressure-reducing valves, pressure-sustaining valves, and pressure-relief valves: locations and how often they are exercised
- blowoffs and air-release valves: locations and how often they are exercised
- distribution system maps: What is shown on maps, how current is the information, and how often is the information updated?

Many utilities survey their distribution networks systematically according to zones, or areas outlined on maps. Other utilities use meter routes because the routes are well thought out and minimize distances in covering the system. However, using meter routes requires familiarity with the system layout.

Leak Detection Equipment

Crew members should have leak detection equipment, including sonic listening equipment with a high-frequency listening probe and a low-frequency ground microphone for pinpointing leaks. When using the ground microphone on turf areas, a "thumb tack" helps provide better-quality sounds. A thumb tack is a flat, metal, horizontal plate attached to a strong, metal, vertical spike. See appendix E to obtain the *AWWA Sourcebook* for leak detection suppliers and consultants. Crew members should also have safety equipment, including safety vests, traffic cones, and barricades.

Tools to measure flow rates should be provided, including a stopwatch, bucket, measuring cup, pressure gauge, and measuring tape or ruler.

Working tools, such as meter-box lid lifters, valve-cover lifters, valve keys, curb-stop keys, small bailing cans or small manual pumps, chalk (keel) or spray paint, pipe locators, and wrenches for tightening meter-spud nuts, should also be provided.

Selecting Team Members

Each leak detection team member should have a keen sense of hearing, ability to discern different sounds, familiarity with water meters and the distribution system, a sense of responsibility, and ability to estimate leak flows, complete leak forms, and work independently.

One person can conduct the initial listening survey, although more may be needed for safety purposes.

Planning the Survey

When planning how the leak survey will be conducted, consider the following:

- What type of noise problems exist within the system?
- What effect will traffic have on the survey?
- What type of protection is required for the leak crew?
- What time of day or night will be most effective to conduct the listening survey?
- What time will be most effective to pinpoint suspected leaks? (Some utilities concentrate on the initial listening phase for several days and pinpoint leaks at the end of the week.)
- Is the crew a compatible group that will work together?
- How will the crew's tasks be divided?
- What is the most effective route to follow in the initial listening survey?
- Which are the most effective leak survey and pinpointing forms, and how are these records to be completed?
- How will leak detection crews communicate and work with repair crews to ensure effectiveness and resolve dry holes?

Leak Detection and Repair Plan

Prepare a leak detection and repair plan. A sample plan is shown in Figure 4-1, and a blank form is included in appendix A.

Team Training

Train team members before conducting a leak detection survey. This will build crew members' confidence and help to ensure that the survey is accurate. Training can be arranged through manufacturers of the equipment to be used—either by attending their training seminars or by on-the-job training as part of the equipment purchase. Training may also be available from consultants, from utilities with existing leak detection programs, or from the state.

Equipment Tune-up

Before conducting the survey, leak detection staff should familiarize themselves with the equipment to be used. Consult the instruction manual and review the instructions. Check the equipment to be sure batteries are charged and worn-out batteries are replaced, electrical and physical connections are tight, and controls work properly. Check to see if the basic leak sound can be heard by testing the equipment on a hose bibb, first with water running and then with no water running.

Have team members practice with the particular pieces of equipment they will use during the survey. All team members should become familiar with the unique background sound of their particular listening instrument; what a quiet sound is (when no leak or water use is present); what water use sounds like (by opening a hose bibb on a customer's service line); the sound when a faucet is dripping; the sound when sprinklers are in use in the area; and the sound when the meter is turning.

LEAK DETECTION PROCEDURES

This section describes, in general terms, how to conduct a leak detection survey. The leak detection daily log (see Figure 4-2) can be used to record the results of the survey, and a blank form is provided in appendix A.

Initial Listening Survey

The objective of the initial listening survey is to listen for leak sounds on contact points in the distribution system. Use the high-frequency contact microphone to listen for leak sounds on all meters, valves, hydrants, blowoffs, air-release valves, and other contact points. Note the address of all locations where water use, meter sounds, or possible leak sounds exist. This initial search through each area of the system can be conducted quickly.

Sound travels a long distance on metallic mains, so listening at contact points allows the listener to hear the sounds of leakage along the length of the main between the points. Sound travels less distance on nonmetallic mains, such as PVC, and a little extra effort may be required during listening surveys. If sound does not carry the entire length of the pipe from one contact point to the next, then the leak detection staff needs to listen over the main itself with a ground microphone.

A number of factors influence how far sound will travel along nonmetallic lines, including system pressure and pipe diameter. The sensitivity of listening equipment also limits the length of pipe along which sounds can be heard.

SAMPLE LEAK DETECTION AND REPAIR PLAN

Name of Agency: County Water Co. Date: 7/18/99

A. Area to Be Surveyed

A-1. Using the results of the water audit, show on a map which areas in the distribution system will be surveyed. Indicate which areas have the higher potential for recoverable leakage. Consider records of previous leaks, type of pipe, age of pipe, soil conditions, high pressures, ground settlement, and improper installation procedures.

Describe each area to be surveyed under item B-2 of this plan.

A-2. Total miles of main to be surveyed: 146

When calculating the miles of main, include the total length of pipe and exclude service lines. If only a portion of the system is being surveyed, calculate the benefit-to-cost ratio to reflect only the portion included in the survey.

A-3. Average number of miles of main surveyed per day: 2.0

The average survey crew can survey about two miles of main per day. Items to consider include distances between services, traffic and safety conditions, and number of listening contact points. Explain if more than three miles per day are surveyed.

A-4. Number of working days needed to complete survey (divide line 2 by line 3): 73

B. Procedures and Equipment

B-1. Describe the procedures and equipment you will use to detect leaks. Experience shows that the best results are obtained by listening for leaks at all system contact points (such as water meters, valves, hydrants, and blowoffs).

Purchase leak detection equipment. Attend manufacturers' training seminars and state training. Conduct initial listening survey on all contact points.

B-2. Describe why the areas noted on the map in step A-1 have the greatest potential for recovering leakage.

Area 1 - Downtown - old ductile-iron mains.

Area 2 - Steel mains over 40 years old.

Area 3 - Remainder of systems.

B-3. If you will not be listening for leaks at all system contact points, describe your plan for effectively detecting leaks.

Not applicable-crew will listen for leaks at all contact points.

B-4. Describe the procedures and equipment you will use to pinpoint the exact location of the detected leaks.

Will use low-frequency ground microphone to listen over pavement surfaces. Consultant's correlator will be used in difficult situations.

B-5. Describe how the leak detection team and the repair crew will work together. How will they resolve the problem of dry holes?

The leak detection crew and the repair crew will jointly excavate all leaks for the first three weeks and resolve any dry holes thereafter.

Figure 4-1 Sample plan for leak detection and repair

B-6. Describe the methods you will use to determine the flow rates for excavated leaks of various sizes.
Bucket-and-stopwatch method will be used on small leaks.
For larger leaks, the diameter of the hole and the pressure will be measured.
The Greeley formula will be used to calculate the volume of large leaks based on measurements taken.

C. Staffing

C-1. How many agency staff will be used? 2

Staff costs including wages and benefits:

	$/hour	$/day
Person 1	17.19	137.52
Person 2	12.50	100.00
TOTAL	29.69	237.52

C-2. How many consultant staff will be used? 1

Cost of consultant staff:

	$/hour	$/day
Person 1	31.25	250.00
Person 2	0	0
TOTAL	31.25	250.00

D. Leak Detection Survey Costs

Leak detection surveys:	$/day	# days	Cost
D-1. Agency-crew costs	237.52	73	17,340
D-2. Consultant-crew costs	250.00	15	3,750
D-3. Vehicle costs	2.40	73	175
D-4. Other			0
D-5. Total survey costs			21,265

E. Leak Detection Budget

E-1. Cost of leak detection equipment	2,550
E-2. Leak detection team training	400
E-3. Leak detection survey costs	21,265
E-4. Total leak detection costs	24,215

F. Leak Survey and Repair Schedule

Indicate realistic, practical dates.

F-1. When will the leak survey begin? August 1, 1999

F-2. When will the leak survey be completed? December 15, 1999

F-3. When will leak repairs begin? August 15, 1999

F-4. When will leak repairs be completed? January 10, 2000

Prepared by:

Name C. M. Biggs Date 7/18/99

Title Manager

Figure 4-1 Sample plan for leak detection and repair (continued)

LEAK DETECTION SURVEY DAILY LOG

Agency: County Water Co. Date 7 / 17 / 99 (Month Day Year)

Leak Detection Team Members: Lloyd Williams, Raymond Goodgion, Wesley Kelso

Manufacturer and Models of Equipment Used: F.S.C. Model L - 100

Area Surveyed: 7 Map Reference: Water Distribution Map

Street and Block Numbers: San Antonio, San Gabriel, San Juan, San Carlos, San Luis, San Miguel 8600 block. Page & Coordinates: ______

Leak Number	Location or Address of Suspected Leak	Agency or Customer (A or C)	Leak Pinpointed (Y or N)	Leak to Be Rechecked (Y or N)	Leak Repaired (Y or N)	Not a Leak (Date)
51	8959 San Antonio	A	Y	N	Y	
52	N/W Cor. Firestone & San Gabriel	A	Y	N	Y	
54	S/W Cor. Firestone & San Gabriel	A	Y	N	Y	
53	S/W Cor. San Juan & Southern	A	Y	N	Y	
55	8990 San Antonio	A	Y	N	Y	
56	8986 San Carlos	A	Y	N	Y	
57	8921 San Luis	A	Y	N	Y	
58	8659 San Miguel	A	Y	N	Y	

	Meters	Hydrants	Valves	Test Rods	Other
Indicate Number of Listening Points Used	483	43	88	0	0

Miles of Main Surveyed 4.242 Survey Time 8 hours

Number of Leaks Suspected 8 To Be Rechecked 8 (number)

Number of Leaks Pinpointed 0 Pinpointing Time 0 hours

Remarks: Found a 50/50 percentage between stem packing leaks and small service meter leaks. Also found two customer sprinkler system leaks, customers were informed.

Figure 4-2 Sample log for leak detection survey

To determine whether it is necessary to listen directly over mains in addition to contact points, perform the following test:

1. Listen over the main with a ground microphone.
2. Have a co-worker turn on a hose bibb at a customer's service.
3. Determine how far away along the main the sound of water escaping from the hose bibb can be heard.

If the distance between contact points is greater than the distance that the sound travels along the main, then use the ground microphone to listen over the main at appropriate intervals between 10 and 50 ft.

Interference. A number of sounds can interfere with leak detection equipment.

Sounds from customer use inside a dwelling include use of showers, toilets, washing machines, pumps, and meters. Even the sound of people talking may be picked up by listening equipment.

Sounds from outside a dwelling can be caused by aircraft, wind and rain, street traffic, interference from power lines or transformers, radio broadcasting, or lawn watering.

Sounds from water noises usually come from adjacent leaks, valves, or turbulence. All of the sounds may be transmitted through leak detection equipment, making it difficult to isolate and identify leak noises.

Faulty equipment, loose electrical connections, improper training, or system pressure less than 15 psi can also obscure or modify leak noises.

Relistening to Suspect Sounds

Return to each location noted in the initial listening survey. Using the high-frequency contact microphone, listen again for the sounds heard earlier. If the location is quiet, there is no leak. If sounds are heard, check to see if the meter is running; a running meter indicates water use. If sounds can still be heard when there is no water use, a leak probably exists. That leak must be pinpointed.

Pinpointing Leaks

The objectives of pinpointing leaks are (1) to determine whether the leak sound is leakage, water use, or some other noise; and (2) to determine the leak's exact location.

Return to the suspected location and again listen for the leak sound. Inspect the area, paying attention to both sight and sound. Use a sonic amplifier, if possible. What might be a leak sound may actually be caused by a pressure-reducing valve, electrical transformer, or other interference.

Review detailed distribution system maps and locate pressure reducers, forgotten valves, or other system apparatus that might make the suspect sound. If when inspecting the area, another possible cause of the sound is found, try to isolate and identify the sound, or quiet it temporarily. For example, a customer pressure-reducing valve can be isolated by shutting off the customer service and then bleeding the pressure off the system by opening the customer's hose bibb. (Be sure to check with the customer before shutting off the service.)

If the leak noise is heard on a water meter (see Figure 4-3), listen carefully for leak sounds on both sides of the meter. Determine if the sound is louder on the customer side or the utility side of the meter. Look for obvious signs of customer use, such as sprinklers operating. In this case, the meter may be heard turning, even if the meter hand is not moving. Then, check the meter indicator for movement; the leak may be in the area of the meter box.

Figure 4-3 Listening for leak noise at meter

If it is difficult to identify which side of the meter the leak is on, notify the customer that the service will be shut off for a few minutes. Close the angle stop and bleed off system pressure from the customer's line by opening the hose bibb. If the leak sound stops, the leak is either within the meter box, on the customer's service line, or in the dwelling. If the noise continues, the leak is on the agency side of the meter.

If the leak is on the customer side of the meter, leave a doorhanger (see Figure 4-4) notifying the customer that there may be a leak in the service line, interior plumbing, or water-using fixtures.

Methods for pinpointing leaks. If a leak is on the main or the service line, the leak sound may be detectable on adjacent service meters, valves, or hydrants. Listen for sounds of leakage on services adjacent to the suspected meter and determine where the sound is the loudest. Pinpointing the exact location can be accomplished by using the ground microphone or the correlator method.

Ground-microphone method. The objective of this method is to find the location of the loudest leak sound over the main or service line.

The first step is to determine the exact location of the main or service. An electronic pipe locator can be used to locate the buried main or service line. Precisely mark the location of the main or service line on the pavement. Locate other nearby pipes from which the sound might be coming.

Ground microphones are either monophonic or stereophonic, depending on the manufacturer. Stereo models can discern differences in intensity between two microphones, but most models have only one microphone.

When using the ground microphone for pinpointing leaks, remember to set the volume relatively low at the beginning, so loud sounds will not be uncomfortable.

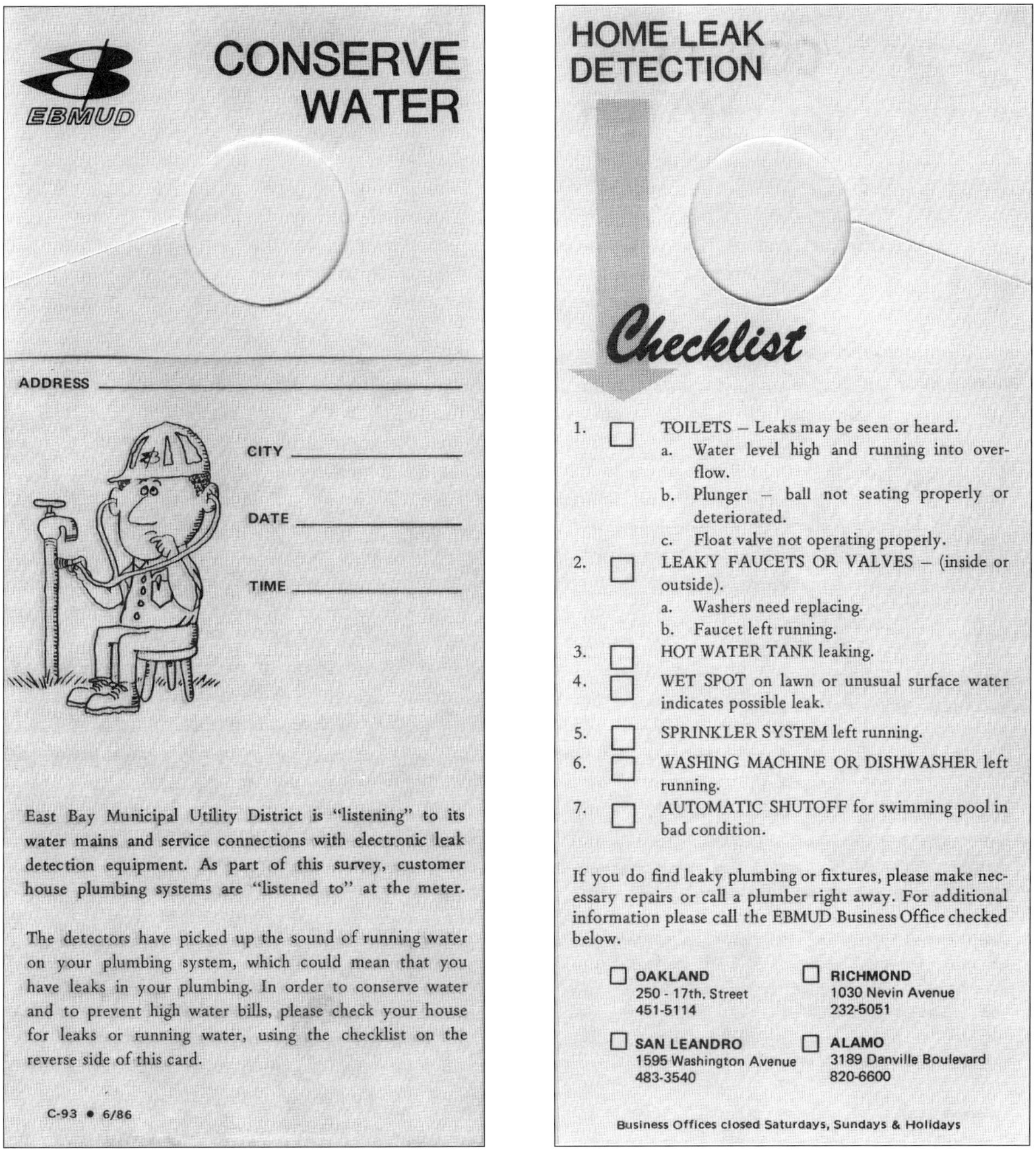

EBMUD

CONSERVE WATER

ADDRESS ____________

CITY ____________

DATE ____________

TIME ____________

East Bay Municipal Utility District is "listening" to its water mains and service connections with electronic leak detection equipment. As part of this survey, customer house plumbing systems are "listened to" at the meter.

The detectors have picked up the sound of running water on your plumbing system, which could mean that you have leaks in your plumbing. In order to conserve water and to prevent high water bills, please check your house for leaks or running water, using the checklist on the reverse side of this card.

C-93 • 6/86

HOME LEAK DETECTION

Checklist

1. ☐ TOILETS – Leaks may be seen or heard.
   a. Water level high and running into overflow.
   b. Plunger – ball not seating properly or deteriorated.
   c. Float valve not operating properly.
2. ☐ LEAKY FAUCETS OR VALVES – (inside or outside).
   a. Washers need replacing.
   b. Faucet left running.
3. ☐ HOT WATER TANK leaking.
4. ☐ WET SPOT on lawn or unusual surface water indicates possible leak.
5. ☐ SPRINKLER SYSTEM left running.
6. ☐ WASHING MACHINE OR DISHWASHER left running.
7. ☐ AUTOMATIC SHUTOFF for swimming pool in bad condition.

If you do find leaky plumbing or fixtures, please make necessary repairs or call a plumber right away. For additional information please call the EBMUD Business Office checked below.

☐ OAKLAND
250 - 17th. Street
451-5114

☐ RICHMOND
1030 Nevin Avenue
232-5051

☐ SAN LEANDRO
1595 Washington Avenue
483-3540

☐ ALAMO
3189 Danville Boulevard
820-6600

Business Offices closed Saturdays, Sundays & Holidays

Courtesy East Bay Municipal Utility District

Figure 4-4 Doorhanger notifying residential customer of possible plumbing leaks

Keep the volume adjustment at the same level throughout each pinpointing sequence. If uncomfortably loud sounds are heard, reduce the volume for safety, then survey the points again to locate the loudest leak sounds.

Use the ground microphone to listen for leak sounds every 5 to 10 ft. Write notes on the sound intensities. If the equipment has a meter, write down meter readings. The strongest signal usually indicates the location of the leak. Be careful not to change the setting of the volume or other controls during this process. Where possible, avoid comparing sounds at points with different surface and compaction characteristics. If this is unavoidable, make allowance for the fact that the same leak sound is quieter at a loosely compacted surface than at a dense one.

After pinpointing the leak, verify its location with a second listening using the ground microphone.

Correlator method. The objective of this method is to pinpoint leak locations using an acoustic leak correlator, or microprocessor, to analyze leak sounds (including those inaudible to the human ear) that travel through the water column and along the pipe wall. These sounds can be picked up, for example, at valves, hydrants, and curb stops. Direct contact with exposed mains or probe rods also can be used.

The correlator method is used in place of or as verification of the previously described ground-microphone method. It is used before drilling or excavating. The correlator method does not rely on the presence of surface sound as does the ground-microphone method. Common noise interference, such as wind, traffic, and ambient system noise, do not affect the leak correlator. The depth of the main, type of cover, and surface conditions are generally not factors to be considered.

To use the leak correlator, the leak sound must be detectable at two or more contact points, and pipe locations and configurations must be accurately determined. The linear pipe distance between the contact points, and pipe materials and diameters are entered into the correlator. Two electronically amplified microphones, connected to and powered by portable electronic preamplifier outstations, are attached to the selected contact points. The leak sound picked up by the microphones and amplified by the outstations is then transmitted to the correlator by a radio housed within the outstation (Figure 4-5).

Courtesy California Department of Water Resources

Figure 4-5 Preamplifier outstation

The leak correlator is essentially a two-channel microprocessor that measures the time delay of a leak noise between two contact points. Although the leak sound's characteristics vary due to such factors as pipe material, diameter, size and nature of the orifice or fissure, system pressures, ground conditions, and other factors, the leak sound velocity (*V*), or speed with which the leak sound travels along the pipe, remains constant.

As previously mentioned, if the sound of the leak can be detected at two contact points, the leak correlator can determine the leak's position. A simple example is shown in Figure 4-6, where the leak is on a main between two sounding points, *A* and *B*, at a distance *D* apart.

The leak is at a point halfway between *C* and *B*. The leak correlator determines the delay in arrival time taken by the leak sound to travel from *C* to *A*, the distance *N*. This delay is the time difference *Td* for the leak sound to reach *A* versus its arrival time at *B*.

Referring to Figure 4-6,

$$D = 2L + N$$

Substituting velocity *V* multiplied by time difference *Td* for *N*

$$D = 2L + VTd$$

The value *D* is measured in the field and velocity *V* is either selected from the leak correlator's memory, or can be computed manually by the operator. The difference in arrival time *Td* of the leak sound at *A* and *B* is automatically established by the correlator through the cross-correlation process. This time the difference is directly related to the sound velocity of the pipe under investigation.

The leak location results appear on the correlator's display, or results can be printed, as shown in Figure 4-7. The operator then measures the indicated distances from the contact points.

While correlators once required permanent vehicle installation, most are now laptop computers with internal rechargable 12-volt direct current (VDC) power supplies (Figure 4-8) and simple to operate.

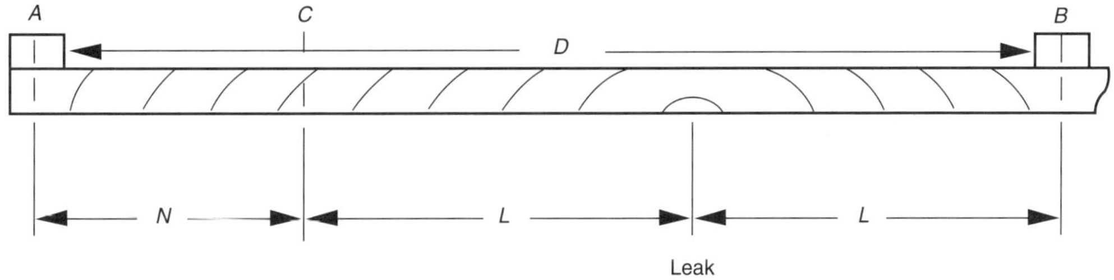

Figure 4-6 Determining the position of a leak using a leak correlator (see text for definitions of variables)

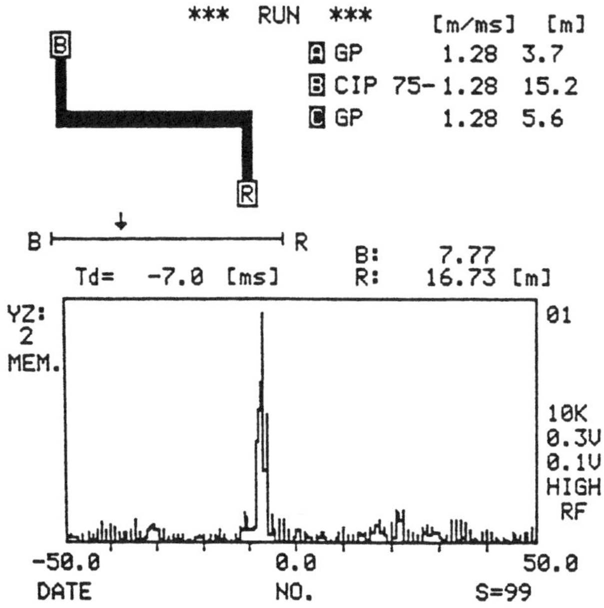

NOTE: The spike on the chart indicates the position of the leak between blue and red transducers.

Figure 4-7 Typical paper-tape output of leak location using correlator

Figure 4-8 Portable laptop leak correlator

A leak correlation system is a package of electronic equipment and accessories, commonly consisting of the following items:

- a laptop microprocessor with internal rechargeable 12-VDC power supply, display screen, internal preamplifier, two-channel internal radio receiver, and stereo headphone
- two electronic amplifier outstations with internal rechargable 12-VDC power supply, internal radio transmitters, microphones, and stereo headphones
- battery charger kit
- manual and test tape with stereo lead

Commonly available accessories include the following:

- cases for carrying items and added protection
- microphone attachment accessory kit
- portable electronic survey tool that serves as a backup outstation
- measuring wheel
- hydrophone sensor package
- stereo recorder with harness
- training tapes
- ground-microphone system
- printer
- pipe locator

Probe method. This is a method to double-check findings using the ground-microphone or correlator method. Drill a small hole through the pavement over the suspected leak taking care not to damage the pipe. Insert a metal rod with a T-handle into the hole and use a high-frequency sonic microphone to listen again for the sound of leakage. Additional holes through the pavement or ground may be drilled as necessary. In unpaved areas, the probe can be used as an extension to listen directly on the buried pipe (see Figure 4-9).

NOTE: For safety and to prevent interruption of service, contact other utilities for clearance before starting to drill. Many areas have a one-call, underground-protection center to clear all utilities at once.

After pinpointing the leak, mark the pavement above the exact location of the leak. Record all information on the leak detection log and turn in work orders for repair.

Nonpersonnel acoustic leak survey. The use of acoustic listening instruments is a proven procedure for identifying and localizing hidden leakage. However, research organizations and practical experience have demonstrated that acoustic listening only on valves and hydrants or the ground surface leads to many leaks being overlooked. Consequently, for effective leakage-reduction programs using acoustic survey, soundings must also be performed on all service connections.

The major disadvantages of this approach include the following factors:

- labor intensive
- high skill levels required
- difficult to maintain efficient performance
- low daily coverage rates
- service line location often difficult and slow
- limited success on nonmetallic pipes

Figure 4-9 Using extension to listen for leak sounds on fire hydrant

Acoustic leak survey results can be optimized by using nighttime operations, uninterrupted listening, and extended listening periods.

Statistical noise analyzer. The statistical noise analyzer makes leakage control possible for areas that cannot be identified as having leaks due to excessive interference. The system includes a hydrophone sensor typically attached at hydrants, measurement taps, or chlorination taps. This sensor is connected by cable to a hand-sized acoustic data processor.

The statistical noise analyzer is designed for nonpersonnel acoustic operations during the night. Because the system listens for a continuous two-hour period at a single contact point, it is more effective at monitoring leak sounds than personnel surveying techniques.

Units are temporarily installed within the distribution system at varying intervals up to approximately 1,500 ft. They are typically preprogrammed to begin listening around 2:00 a.m. When activated, the system monitors noise measured on the distribution system and stores results in twenty-four 5-min analysis periods until monitoring is complete (4:00 a.m.).

The statistical variance of this noise is determined by the presence or absence of leakage. The "noise signature" obtained at each monitoring point confirms the presence or absence of leakage and indicates the relative location.

Installation and data collection is usually performed during daily operations. The data can be downloaded in the field for immediate or later evaluation by technicians or engineering staff. The unit can then be reprogrammed and installed at a new location for the next night's sounding operation.

Most software provides an overall 2-h graph, a 3-D graph, and 24 individual 5-min graphs.

This method of leakage identification and location includes the following benefits:

- Low deployment cost. The unit can be installed during daytime hours and picked up the following day for positioning in other areas.
- Minimal training to set up. Deployment is simple and can be performed with little training.
- Easy data collection. Interrogation is by personal computer and can be performed in the field, or data can be stored for later analysis by appropriate technical personnel.
- High acoustic sensitivity. Hydrophone sensors are used to directly contact the pressurized water.
- Nighttime listening. Provides the benefits of nighttime listening without the labor costs.
- Unlike personnel listening, there is no timing of the listening step, concentration and distractions are not factors, and trained skills are more easily transferred.

When the system is installed in visible areas, security and equipment protection is maintained by placing the units in locked security housings.

In addition to progressive acoustic leak survey operations, the units can also be deployed for

- specific areas of difficulty for extensive analysis
- advanced known leak investigation
- extended daily monitoring of special event or special concern areas
- investigation of piping sections that have had repairs
- inspection of new installations
- permanent monitoring of trunk mains or any piping section where daily monitoring is desirable

The statistical noise analyzer's application lends itself to distribution system operations from small rural systems to large municipalities, as well as industrial and commercial water systems.

Tracer gas method. Occasionally situations occur where leaks cannot be detected or pinpointed by traditional electrosonic or correlation methods. These types of leaks often occur as hydrostatic test failures during construction. They are usually small and occur more frequently with nonmetallic pipe materials. Tracer gas has proven effective for detecting and pinpointing leaks in these situations.

The tracer gas method uses two gases: helium and hydrogen. For helium detection, the method involves dewatering the section of main or pipe being tested and injecting a gas mixture of 5 to 10 percent helium (balance air) at one end of the section. A relief is kept open at the opposite end to allow the helium to flow through and fill the test section. When helium is detected at the relief end, the relief is closed. The section is then pressurized to a predetermined pressure.

For detection using hydrogen gas, it is not necessary to dewater the main because the mixture (less than 5 percent) is injected in a liquid form into the water. The gas mixture is a standard mixture of 5 percent hydrogen in nitrogen, purchased already mixed from a gas supplier. **CAUTION: the actual blending of hydrogen and nitrogen is a highly hazardous operation that should only be undertaken by the gas supplier.** Do not handle hydrogen gas in any form other than ready-mixed to 5 percent hydrogen in nitrogen, or less. Any hydrogen–nitrogen mixture containing less than 5.7 percent hydrogen is nonflammable (ISO 10156).

As the liquid exits the leak, it returns to a gaseous form. Walking directly over the test section of pipe, the operator uses a specialized instrument that continuously senses the atmosphere at grade. The instrument is highly sensitive and can detect minor seepages of gas to atmosphere.

When gas is detected at the surface, the instrument's various sensitivity settings can quickly verify and pinpoint the leak location. If the surface over the pipe is covered with asphalt or concrete, or soil conditions include frost, it may be necessary to place test holes directly over the pipe, normally at 10-ft intervals along the pipe run, to allow the helium to vent to atmosphere through the cover.

Excavating the leak. The survey crew and the repair crew should work together to uncover the leak. If the hole is dry, the survey crew can relisten to the sound and help the repair crew locate the leak. A leak may be missed because it is on the bottom of the pipe or a few inches away and no sign of dampness or water is visible. By working together, both the survey and repair crews can share knowledge and experience that make locating the leak easier.

Uncovering leaks requires careful excavation to avoid other pipes, other utilities, or both. Be sure to check with other utilities before starting to dig.

Table 4-1 Drips per second converted to gallons per minute

Drips per Second	Gallons per Minute
1	0.006
2	0.012
3	0.018
4	0.024
5	0.030

NOTE: Five drips per second amounts to a steady stream.

Table 4-2 Cups per minute converted to gallons per minute

8-oz Cups per Minute	Gallons per Minute
0.25	0.016
0.50	0.031
0.75	0.047
1.00	0.062
1.50	0.094
2.00	0.125
2.50	0.156
3.00	0.188
3.50	0.219
4.00	0.250

MEASURING AND ESTIMATING LOSSES FROM DISCOVERED LEAKS

As part of detection and repair, leaks should be measured to determine the rate of loss and the total volume lost during the life of the leak. There are three ways to do this: use a container of known volume and a stopwatch, use a hose and a meter, or calculate losses using modified-orifice and friction-loss formulas.*

Bucket-and-Stopwatch Method

The bucket-and-stopwatch method is as simple as its name.

Hold a container against the leak for a predetermined time period. Measure the time with a stopwatch. Measure the water captured with a measuring cup or other container of known volume. Then convert time and volume to gallons per minute (see Tables 4-1 and 4-2).

Use time intervals that are easy to deal with. Catch the leaking water for 1 min, then the volume collected is the per-minute flow. For other time periods, see the following:

Time in seconds:	6	10	15	30	
Multiply volume in gallons by:	10	6	4	2	to get gallons per minute.

Table 4-2 provides the conversion from cups per minute to gallons per minute. To convert gallons per minute to million gallons for a 2-year time period (the average lifetime of a leak), use the following:

$$\text{A leak of 1.0 gpm for 2 years} = \frac{(60\ \text{min/h})(24\ \text{h/day})(365\ \text{days/year})(2\ \text{years})}{1{,}000{,}000\ \text{gal}}$$

$$= 1.051\ \text{mil gal} \qquad \text{(Eq 4-1)}$$

Large, spraying leaks can be measured by draping an enveloping device (such as a large canvas, rain jacket, or inverted pail) over the leak and diverting the water into a container.

Hose-and-Meter Method

This is the most direct method of measuring leaks, but it requires some mechanical effort. Connect a hose to the leak and direct the flow through a meter. Then, simply read the meter.

Calculation Method

This is the simplest method to perform in the field, but it requires calculations. The method is often helpful for large leaks where the flow is too great to measure and the main must be valved off. It requires measuring the size and shape of the hole and determining the line pressure. A pressure gauge or a hand-held pitot blade could be used to determine the pressure of the water coming from the leak or a nearby fire hydrant. This method also uses some assumptions regarding the shape of the hole, which may introduce error.

---

*Greeley, D.S. "Leak Detection Productivity," *Reference Number 1981,* Water/Emergency & Management, Des Plaines, Ill. (1981).

Table 4-3 Leak losses for circular holes under different pressures*

Diameter of Hole *in.*	Area of Hole *in.*^{2}	Leak Losses—*gpm* Water Pressure—*psi* 20	40	60	80	100	120	140	160	180	200
0.1	0.007	1.067	1.510	1.850	2.136	2.388	2.616	2.825	3.021	3.204	3.337
0.2	0.031	4.271	6.041	7.399	8.544	9.522	10.464	11.302	12.083	12.816	13.509
0.3	0.070	9.611	13.593	16.648	19.224	21.493	23.544	25.430	27.186	28.835	30.395
0.4	0.125	17.087	24.165	29.597	34.175	38.209	41.856	45.209	48.331	51.263	54.036
0.5	0.196	26.699	37.758	46.245	53.399	59.702	65.400	70.640	75.518	80.098	84.431
0.6	0.282	38.477	54.372	66.593	76.894	85.971	94.176	101.721	108.745	115.341	121.581
0.7	0.384	52.331	74.007	90.640	104.662	117.010	128.184	138.454	148.014	156.993	165.485
0.8	0.502	68.350	96.662	118.387	136.701	152.840	167.424	180.839	193.325	205.052	216.144
0.9	0.636	86.506	122.338	149.833	173.012	193.434	211.896	228.874	244.676	259.519	273.557
1.0	0.785	106.798	151.035	184.979	213.596	238.807	261.600	282.561	302.070	320.394	337.725
1.1	0.950	129.225	182.752	223.825	258.451	288.957	316.536	341.898	365.505	387.676	408.647
1.2	1.131	153.789	217.490	266.370	307.578	343.882	376.704	406.887	434.981	461.367	486.323
1.3	1.327	180.488	255.249	312.615	360.977	403.584	442.104	477.527	510.498	541.465	570.755
1.4	1.539	209.324	296.028	362.559	418.648	468.062	512.737	553.819	592.057	627.972	661.941
1.5	1.767	240.295	339.829	416.203	480.590	537.317	588.601	635.762	679.658	720.886	759.880
1.6	2.011	273.402	386.649	473.547	546.805	611.347	669.697	723.355	773.299	820.208	864.575
1.7	2.270	308.646	436.491	534.590	617.292	690.153	756.025	816.600	872.983	925.938	976.024
1.8	2.545	346.025	489.353	599.333	692.050	773.736	847.585	915.496	978.707	1,038.070	1,094.220
1.9	2.836	385.540	545.237	667.776	771.081	862.095	944.378	1,020.040	1,090.470	1,156.620	1,219.180
2.0	3.142	427.191	604.140	739.918	854.383	955.230	1,046.400	1,130.240	1,208.280	1,281.570	1,350.890

*Calculated using Greeley's formula (see Eq 4-2).

Table 4-4 Leak losses for joints and cracks*

Area of Joint or Crack Length *in.*	Width *in.*	Leak Losses—*gpm* Water Pressure—*psi* 20	40	60	80	100	120	140	160	180	200
1.0	1/32	3.2	4.5	5.5	6.4	7.1	7.8	8.4	9.0	9.6	10.1
1.0	1/16	6.4	9.0	11.0	12.7	14.2	15.6	16.9	18.0	19.1	20.1
1.0	1/8	12.7	18.0	22.1	25.5	28.5	31.2	33.7	36.0	38.2	40.3
1.0	1/4	25.5	36.0	44.1	51.0	57.0	62.4	67.4	72.1	76.5	80.6

*For leaks emitted from joints and cracked service pipes, an orifice coefficient of 0.60 is used in the following equation:

$$Q = (22.796)(A)(\sqrt{P})$$

Where: Q = flow, in gallons per minute; A = the area, in square inches; P = the pressure, in pounds per square inch.

For losses from such items as pipes or broken taps, assume an orifice coefficient of 0.80 and calculate flow in gallons per minute from the formula

$$Q = \frac{43{,}767}{1{,}440} \times A \times \sqrt{P} \qquad \text{(Eq 4-2)}$$

Where:

Q = flow, in gallons per minute
A = the cross-sectional area of the leak, in square inches
P = pressure, in pounds per square inch

If a hole in a pipe were circular, then the area would be $A = 3.14\ r^2$. Measure the diameter of the hole (to get the radius) and determine the pressure in the pipe.

LEAK REPAIR REPORT

Agency: County Water Co. Date 11 / 5 / 99 (Month Day Year)

W.O. No.: 10077 Foreman: Hal Nielson

LEAK IDENTIFICATION Map Reference: Water Distribution System

Refer to Leak Discovery Report Page and Coordinates: Area 13

Discovery Date: 11/4/99 Leak No.: 197

Location (include street name and number): 9224 Garden View

FOR MAIN AND SERVICE LATERAL LEAKS ONLY

Sketch a map of the site including:

1. Street name; north arrow.
2. Meter number (if applicable).
3. Mains and hydrants in shutdown area.
4. All valves (give valve numbers and show which were closed during repair).
5. Locate leak to nearest intersection or house with address. Show distances to property lines or street centerlines.

If Main or Service Leak, Attach Three Photos:

1. Straight down over leak or damage.
2. Close-up of leak and damage.
3. Any other photo which you feel will help.

Leak Found? Yes (Yes/No)

TYPE OF LEAK

Meter Leak	____	Main Line Leak	____	Joint Leak	____
Meter Spud Leak	____	Service Lateral Leak	____	Other Leak	____
Meter Yoke Leak	____	Fire Hydrant Leak	____	Describe	automatic
Curb Stop Leak	____	Valve Leak	____		flushing device

DESCRIPTION OF REPAIR

Damaged part was: X Repaired ____ Replaced

If replaced, what material was used? 3 bags AC patch

If repaired, what repairs were made?

____ Leak Clamp ____ Repacked Valve

____ Welded ____ Recaulked Joint

____ Other (describe) ____

Repair Time 4 h (From/To)

Crew Size 2 (persons)

Equipment Used for Repair

____ Backhoe

____ Dumptruck

Repair Costs:

Materials	$ 11.19
Labor	$ 114.72
Equipment	$ 12.00
Other	$ 1.00
Total	$ 138.91

Size of Leak:

Measured 3½ gpm

Estimated ____ gpm

Method Used: Timed 3½ gallon bucket with stopwatch

Figure 4-10 Leak repair report

DESCRIPTION OF DAMAGE FOR MAINS AND SERVICES

What part was damaged?

_____ Pipe Barrel _____ Flange Nuts, Bolts, Tie Rods

_____ Joint _____ Other (describe)

_____ Valve ____________________

In your opinion, what caused the damage? ________

____

____

Estimated Age of Leak in Months 48

How Determined ____________________

Diameter of Main or Lateral in Inches __________

Depth to Top of Pipe in Inches __________

Type of Break

_____ Split

_____ Hole

_____ Circumferential Split

_____ Broken Coupling

_____ Service Pulled

_____ Cracked at Corporation Stop

_____ Gasket Blown

_____ Crushed Pipe

_____ Cracked Bell

_____ Other (describe) ____________________

____

Pipe Material:

_____ Galv. Iron _____ Ductile Iron _____ A.C.P.

_____ Black Iron _____ Steel _____ P.V.C.

_____ Cast Iron _____ Copper _____ Polybutylene

System Pressure ____________________

How Determined ____________________

____

Examine broken edge of cast- or ductile-iron pipe:

Original Thickness: __________ Inches

Min. Thickness of Good Grey Metal Remaining: __________ Inches

Deterioration is on: _____ Outside _____ Inside

Is there evidence of previous leak or repairs in same general area? _____ Yes _____ No

Last Repair Date (if known) ____________________

Number of Previous Leak Repair Clamps Present ____________________

Cause of Leak ____________________

____

In your opinion, should pipe be replaced? _____ Yes _____ No _____ Do not know

If yes, explain extent: ____________________

____

FOR EXCAVATIONS, INDICATE GROUND CONDITIONS

Type of Soil:

_____ Rocky _____ Sandy

_____ Clay _____ Hard Pan

_____ Adobe X Loam

_____ Other ____________

Existing Bedding:

_____ Gravel/Sand

X Native Soil

_____ Pea Gravel

_____ Other ____________

Type of Cover:

_____ Concrete

_____ Asphalt

_____ Soil

_____ Other ____________

Figure 4-10 Leak repair report (continued)

LEAK DETECTION AND REPAIR PROJECT SUMMARY

Agency: County Water Company

Name of Report Preparer: C.M. Biggs

Date: 12/15/99

LEAK DETECTION SURVEY

Total Number of Days Leak Surveys Were Conducted: 65

First Survey Date: 7/10/99 Last Survey Date: 11/4/99

	Meters	Hydrants	Valves	Test Rods	Other
Number of Listening points:	13,786	1,067	2,505	0	17

Number of suspected leaks: 175 Number of pinpointed leaks: 7

Survey time: 469 hours Miles of main surveyed: 124.7

Pinpointing time: 2 hours

$$\text{Average survey rate} = \frac{\text{miles of main surveyed} \times 8}{\text{total survey and pinpointing hours}} = 2.1 \text{ miles per day}$$

Total number of visible leaks reported since survey started, from other sources (not discovered during leak detection surveys): 0

LEAK REPAIR SUMMARY

First Leak Repair Made: 7/11/99 Last Leak Repair Made: 11/5/99

Number of Repairs Needing Excavation: 27 Number of Repairs Not Needing Excavation: 148 Total Number of Repaired Leaks: 175

Total Water Losses From Excavated Leaks: 203.5 gpm Total Water Losses From Nonexcavated Leaks: 2.9 gpm Total Water Losses: 206.4 gpm

	Excavated Leak Repair Costs		Nonexcavated Leak Repair Costs		Total Repair Costs
Materials	$ 699.36	Materials	$ 411.68	Materials	$ 1,111.04
Labor	$ 4,377.39	Labor	$ 2,255.72	Labor	$ 6,633.11
Equipment	$ 561.40	Equipment	$ 248.75	Equipment	$ 810.15
Other	$ 35.00	Other	$ 83.50	Other	$ 118.50
Subtotal	$ 5,673.15	Subtoal	$ 2,999.65	Total	$ 8,672.80

Figure 4-11 Leak detection and repair project summary

LEAK DETECTION PROJECT COST-EFFECTIVENESS

Step 1. Calculate the value of water recovered (Vwr) from all repaired leaks.

(Vwr) = (total leakage recovered in gpm)(water cost, Wc)

Wc = Line 19 of Water Audit Report = water purchase price + operating costs per unit of water

Vwr = 206.4 gpm × 60 min × 24 hrs × 730 days × $ 843 /mil gal = $ 182,904

Step 2. Determine the total cost of the leak detection survey.

Leak Detection Survey Costs

Equipment	$ 2,404.65
Training	$ 400.00
Survey Costs	$ 25,944.84
Total	$ 28,749.49

Step 3. Divide Vwr (from step 1) by the total costs (calculated in step 2).

$$\text{Benefit:Cost Ratio (B:C)} = \frac{\text{value of water recovered}}{\text{total cost leak detection survey}}$$

= 6.4

For planning future leak detection efforts, you can calculate average survey costs per mile.

Step 4. Determine average survey costs per mile of main surveyed (C/mi).

$$\text{C/mi} = \frac{\text{total cost of leak detection survey}}{\text{total number of miles surveyed}} = \frac{\$\,28{,}749.00}{124.70\text{ mi}}$$

C/mi = $ 230.6 /mi

Figure 4-11 Leak detection and repair project summary (continued)

For relatively small holes, leak rates are calculated by assuming a circular hole and several pressures. Tables 4-1 and 4-2 show calculated leak rates for typical meter-box leaks. Table 4-3 covers circular leaks, and Table 4-4 covers joints and cracks.

DETERMINING LEAK DETECTION EFFECTIVENESS

After repairing leaks, record all information regarding excavation, flow rates, and repair on the leak repair report (see Figure 4-10). This helps with future repair projects and provides information to be used in evaluating benefits of the leak-detection project.

An important and often neglected postsurvey step is determining whether or not the project was a cost-effective water conservation measure. To determine if it was cost-effective, the agency must evaluate the completed leak detection project.

The Leak Detection and Repair Project Summary (see Figure 4-11) includes information needed for this evaluation. Blank forms are provided in appendix A.

Appendix A

Blank Forms

This Appendix includes blank forms for use in a water audit and leak detection and repair program. Forms include

- water audit worksheet
- leak detection and repair plan
- leak detection survey daily log
- leak repair report
- leak detection and repair project summary

Instructions for completing the water audit worksheet are given in chapter 2. Instructions for completing all other forms are discussed in chapter 4.

WATER AUDIT WORKSHEET

For: ______________________ Audit Study Period: ______________________

Line	Item	Water Volume: Subtotal	Water Volume: Total Cumulative	Units*
Task 1—Measure the Supply				
1	Uncorrected total water supply to the distribution system (total of master meters)		______	______
2A–C	Adjustments to total water supply			
2A	Source meter error (+ or –)	______		______
2B	Change in reservoir and tank storage (+ or –)	______		______
2C	Other contributions or losses (+ or –)	______		______
3	Total adjustments to total water supply (add lines 2A, 2B, and 2C)		______	______
4	Adjusted total water supply to the distribution system (add lines 1 and 3)		______	______
Task 2—Measure Authorized Metered Use				
5	Uncorrected total metered water use	______		______
6	Adjustments due to meter reading lag time (+ or –)	______		______
7	Metered deliveries (add lines 5 and 6)		______	______
8A–C	Total sales meter error and system-service meter errors (+ or –)			
8A	Residential meter error	______		______
8B	Large meter error	______		______
8C	Total (add lines 8A and 8B)		______	______
9	Corrected total metered water deliveries (add lines 7 and 8C)		______	______
10	Corrected total unmetered water (subtract line 9 from line 4)		______	______
Task 3—Measure Authorized Unmetered Use				
11A	Firefighting and firefighting training	______		______
11B	Main flushing	______		______

NOTE: 1 acre-ft = 43,560 ft^3 = 325,851 gal.

*Units of measure must be consistent throughout the worksheet. The particular unit used (that is, acre-feet, millions of gallons, cubic feet, cubic metres, or other unit) is left to the user.

Form continues on next page.

Line	Item	Water Volume Subtotal	Total Cumulative	Units*
	Measure authorized unmetered use (continued)			
11C	Storm-drain flushing	_____		_____
11D	Sewer cleaning	_____		_____
11E	Street cleaning	_____		_____
11F	Schools	_____		_____
11G	Landscaping in large public areas:			
	Parks	_____		_____
	Golf courses	_____		_____
	Cemeteries	_____		_____
	Playgrounds	_____		_____
	Highway median strips	_____		_____
	Other landscaping	_____		_____
11H	Decorative water facilities	_____		_____
11I	Swimming pools	_____		_____
11J	Construction sites	_____		_____
11K	Water quality and other testing (pressure-testing pipe, water quality, etc.)	_____		_____
11L	Process water at treatment plants	_____		_____
11M	Other unmetered uses	_____		_____
12	Total authorized unmetered water (add lines 11A through 11M)		_____	_____
13	Total water losses (subtract line 12 from line 10)		_____	_____
Task 4—Measure Water Losses				
14A	Accounting procedure errors	_____		_____
14B	Unauthorized connections	_____		_____
14C	Malfunctioning distribution system controls	_____		_____
14D	Reservoir seepage and leakage	_____		_____
14E	Evaporation	_____		_____

NOTE: 1 acre-ft = 43,560 ft^3 = 325,851 gal.

*Units of measure must be consistent throughout the worksheet. The particular unit used (that is, acre-feet, millions of gallons, cubic feet, cubic metres, or other unit) is left to the user.

Form continues on next page.

Line	Item	Water Volume Subtotal	Total Cumulative	Units*
	Measure identified water losses (continued)			
14F	Reservoir overflow	________		________
14G	Discovered leaks	________		________
14H	Unauthorized use	________		________
15	Total identified water losses (add lines 14A through 14H)		________	________
Task 5—Analyze Audit Results				
16	Potential water system leakage (subtract line 15 from line 13)		________	________
17	Recoverable leakage (multiply line 16 by 0.50)		________	________

Line	Item	Dollars per Unit of Volume
18A–B	Cost savings	
18A	Cost of water supply	____________
18B	Variable operation and maintenance costs	____________
19	Total costs per unit of recoverable leakage (add lines 18A and 18B)	____________

Line	Item	Dollars per Year
20	One-year benefit from recoverable leakage (multiply line 17 by line 19)	____________
21	Total benefits from recovered leakage (multiply line 20 by 2)	____________
22	Total costs of leak detection project	____________
23	Benefit-to-cost ratio (divide line 21 by line 22)	____________

Prepared by:

Name ________________________________

Title ________________________________ Date ____________

NOTE: 1 acre-ft = 43,560 ft^3 = 325,851 gal.

*Units of measure must be consistent throughout the worksheet. The particular unit used (that is, acre-feet, millions of gallons, cubic feet, cubic metres, or other unit) is left to the user.

LEAK DETECTION AND REPAIR PLAN

Name of Agency: ______________________________ Date: ______________

A. Area to be Surveyed

A-1. Using the results of the water audit, show on a map which areas in the distribution system will be surveyed. Indicate which areas have the higher potential for recoverable leakage. Consider records of previous leaks, type of pipe, age of pipe, soil conditions, high pressures, ground settlement, and improper installation procedures.

Describe each area to be surveyed under item B-2 of this plan.

A-2. Total miles of main to be surveyed: ______________

When calculating the miles of main, include the total length of pipe and exclude service lines. If only a portion of the system is being surveyed, calculate the benefit-to-cost ratio to reflect only the portion included in the survey.

A-3. Average number of miles of main surveyed per day: ______________

The average survey crew can survey about two miles of main per day. Items to consider include distances between services, traffic and safety conditions, and number of listening contact points. Explain if more than three miles per day are surveyed. ______________________________

A-4. Number of working days needed to complete survey (divide line 2 by line 3): ______________

B. Procedures and Equipment

B-1. Describe the procedures and equipment you will use to detect leaks. Experience shows that the best results are obtained by listening for leaks at all system contact points (such as water meters, valves, hydrants, and blowoffs). ______________________________

B-2. Describe why the areas noted on the map in step A-1 have the greatest potential for recovering leakage.

B-3. If you will not be listening for leaks at all system contact points, describe your plan for effectively detecting leaks. ______________________________

B-4. Describe the procedures and equipment you will use to pinpoint the exact location of the detected leaks.

B-5. Describe how the leak detection team and the repair crew will work together. How will they resolve the problem of dry holes? ______________________________

Form continues on next page.

B-6. Describe the methods you will use to determine the flow rates for excavated leaks of various sizes.

____

____

____

____

C. Staffing

C-1. How many agency staff will be used? __________

Staff costs including wages and benefits:

Person 1	$/hour	________	$/day	________
Person 2	$/hour	________	$/day	________
TOTAL	$/hour	________	$/day	________

C-2. How many consultant staff will be used? __________

Cost of consultant staff:

Person 1	$/hour	________	$/day	________
Person 2	$/hour	________	$/day	________
TOTAL	$/hour	________	$/day	________

D. Leak Detection Survey Costs

Leak detection surveys:	$/day	# days	Cost
D-1. Agency-crew costs	________	________	________
D-2. Consultant-crew costs	________	________	________
D-3. Vehicle costs	________	________	________
D-4. Other	________	________	________
D-5. Total survey costs	________	________	________

E. Leak Detection Budget

E-1. Cost of leak detection equipment	________
E-2. Leak detection team training	________
E-3. Leak detection survey costs	________
E-4. Total leak detection costs	________

F. Leak Survey and Repair Schedule

Indicate realistic, practical dates.

F-1. When will the leak survey begin? ____

F-2. When will the leak survey be completed? ________________________________

F-3. When will leak repairs begin? ____

F-4. When will leak repairs be completed? ________________________________

Prepared by:

Name ________________________________ Date ________________

Title ________________________________

LEAK DETECTION SURVEY DAILY LOG

Agency: ______________________________ Date ______________

Leak Detection Team Members: ______________________________

Manufacturer and Models of Equipment Used: ______________________________

Area Surveyed: ______________ Map Reference: ______________

Street and Block Numbers: ______________________ Page & Coordinates: __________

Leak Number	Location or Address of Suspected Leak	Agency or Customer (A or C)	Leak Pinpointed (Y or N)	Leak to Be Rechecked (Y or N)	Leak Repaired (Y or N)	Not a Leak (Date)

	Meters	Hydrants	Valves	Test Rods	Other
Indicate Number of Listening Points Used	______	______	______	______	______

Miles of Main Surveyed ______________ Survey Time __________ hours

Number of Leaks Suspected ______________ To Be Rechecked __________ (number)

Number of Leaks Pinpointed ______________ Pinpointing Time __________ hours

Remarks: ____

LEAK REPAIR REPORT

Agency: ______________________ Date ______________

W.O. No.: ______________ Foreman: ______________

LEAK IDENTIFICATION Map Reference: ______________

Refer to Leak Discovery Report Page and Coordinates: ______________

Discovery Date: ______________ Leak No.: ______________

Location (include street name and number): ______________

FOR MAIN AND SERVICE LATERAL LEAKS ONLY

Sketch a map of the site including:

1. Street name; north arrow.
2. Meter number (if applicable).
3. Mains and hydrants in shutdown area.
4. All valves (give valve numbers and show which were closed during repair).
5. Locate leak to nearest intersection or house with address. Show distances to property lines or street centerlines.

If Main or Service Leak, Attach Three Photos:

1. Straight down over leak or damage.
2. Close-up of leak and damage.
3. Any other photo which you feel will help.

Leak Found? ______________ (Yes/No)

TYPE OF LEAK

Meter Leak	_____	Main Line Leak	_____	Joint Leak	_____
Meter Spud Leak	_____	Service Lateral Leak	_____	Other Leak	_____
Meter Yoke Leak	_____	Fire Hydrant Leak	_____	Describe	______________
Curb Stop Leak	_____	Valve Leak	_____		______________

DESCRIPTION OF REPAIR

Damaged part was: _____ Repaired _____ Replaced

If replaced, what material was used? ______________

____

If repaired, what repairs were made?

_____ Leak Clamp _____ Repacked Valve

_____ Welded _____ Recaulked Joint

_____ Other (describe) ______________

____

Repair Time ______________ (From/To)

Crew Size ______________ (persons)

Equipment Used for Repair

_____ Backhoe

_____ Dumptruck

Repair Costs:

Materials $ ______________

Labor $ ______________

Equipment $ ______________

Other $ ______________

Total $ ______________

Size of Leak:

Measured ______________ gpm

Estimated ______________ gpm

Method Used: ______________

____

Form continues on next page.

DESCRIPTION OF DAMAGE FOR MAINS AND SERVICES

What part was damaged?

_____ Pipe Barrel _____ Flange Nuts, Bolts, Tie Rods

_____ Joint _____ Other (describe)

_____ Valve ____________________________

In your opinion, what caused the damage? ________

____

____

Estimated Age of Leak in Months ________________

How Determined ____________________________

Diameter of Main or Lateral in Inches _____________

Depth to Top of Pipe in Inches _________________

Type of Break

_____ Split

_____ Hole

_____ Circumferential Split

_____ Broken Coupling

_____ Service Pulled

_____ Cracked at Corporation Stop

_____ Gasket Blown

_____ Crushed Pipe

_____ Cracked Bell

_____ Other (describe) __________________________

____

Pipe Material:

_____ Galv. Iron _____ Ductile Iron _____ A.C.P.

_____ Black Iron _____ Steel _____ P.V.C.

_____ Cast Iron _____ Copper _____ Polybutylene

System Pressure __________________________

How Determined __________________________

____

Examine broken edge of cast- or ductile-iron pipe:

Original Thickness: ________________ Inches

Min. Thickness of Good Grey Metal Remaining: ________________ Inches

Deterioration is on: _____ Outside _____ Inside

Is there evidence of previous leak or repairs in same general area? _____ Yes _____ No

Last Repair Date (if known) ______________________

Number of Previous Leak Repair Clamps Present ____________________

Cause of Leak __________________________________

____

In your opinion, should pipe be replaced? _____ Yes _____ No _____ Do not know

If yes, explain extent: ____

____

FOR EXCAVATIONS, INDICATE GROUND CONDITIONS

Type of Soil:

_____ Rocky _____ Sandy

_____ Clay _____ Hard Pan

_____ Adobe _____ Loam

_____ Other _________________

Existing Bedding:

_____ Gravel/Sand

_____ Native Soil

_____ Pea Gravel

_____ Other _________________

Type of Cover:

_____ Concrete

_____ Asphalt

_____ Soil

_____ Other _________________

LEAK DETECTION AND REPAIR PROJECT SUMMARY

Agency: ______________________________

Name of Report Preparer: ______________________________

Date: ______________________

LEAK DETECTION SURVEY

Total Number of Days Leak Surveys Were Conducted: __________

First Survey Date: ____________________ Last Survey Date: ____________________

	Meters	Hydrants	Valves	Test Rods	Other
Number of Listening points:	______	______	______	______	______

Number of suspected leaks: __________ Number of pinpointed leaks: __________

Survey time: __________ hours Miles of main surveyed: __________

Pinpointing time: __________ hours

$$\text{Average survey rate} = \frac{\text{miles of main surveyed} \times 8}{\text{total survey and pinpointing hours}} = \text{\_\_\_\_\_\_ miles per day}$$

Total number of visible leaks reported since survey started, from other sources (not discovered during leak detection surveys): ________

LEAK REPAIR SUMMARY

First Leak Repair Made: ____________________ Last Leak Repair Made: ____________________

Number of Repairs Needing Excavation: ______ Number of Repairs Not Needing Excavation: ______ Total Number of Repaired Leaks: ______

Total Water Losses From Excavated Leaks: _____ gpm Total Water Losses From Nonexcavated Leaks: _____ gpm Total Water Losses: ______ gpm

Excavated Leak Repair Costs		Nonexcavated Leak Repair Costs		Total Repair Costs	
Materials	$ ________	Materials	$ ________	Materials	$ ________
Labor	$ ________	Labor	$ ________	Labor	$ ________
Equipment	$ ________	Equipment	$ ________	Equipment	$ ________
Other	$ ________	Other	$ ________	Other	$ ________
Subtotal	$ ________	Subtoal	$ ________	Total	$ ________

Form continues on next page.

LEAK DETECTION PROJECT COST-EFFECTIVENESS

Step 1. Calculate the value of water recovered (Vwr) from all repaired leaks.

(Vwr) = (total leakage recovered in gpm)(water cost, Wc)

Wc = Line 19 of Water Audit Report = water purchase price + operating costs per unit of water

Vwr = _______ gpm × 60 min × 24 hrs × ______ days × $ ______ /mil gal = $ _______________

Step 2. Determine the total cost of the leak detection survey.

Leak Detection Survey Costs

Equipment	$ ___________________
Training	$ ___________________
Survey Costs	$ ___________________
Total	$ ___________________

Step 3. Divide Vwr (from step 1) by the total costs (calculated in step 2).

$$\text{Benefit:Cost Ratio (B:C)} = \frac{\text{value of water recovered}}{\text{total cost leak detection survey}}$$

= __________

For planning future leak detection efforts, you can calculate average survey costs per mile.

Step 4. Determine average survey costs per mile of main surveyed (C/mi).

$$\text{C/mi} = \frac{\text{total cost of leak detection survey}}{\text{total number of miles surveyed}} = \frac{\$\ \_\_\_\_\_\_\_\_\_\_}{\_\_\_\_\_\_\_\_\_\_\ \text{mi}}$$

C/mi = $ _____________ /mi

Appendix B

Meters: Types, Uses, Sizing, and Installation

Measuring accurately is what meters are designed to do. But there is more to meter accuracy than calibration: The type and size of meter, and how it is installed, also affect its accuracy. That is why a meter may pass a bench test but fail in the field. This appendix describes the various types of meters and their uses, then discusses sizing and installation.

TYPES OF METERS

Meters can be classified into five general categories, based on the way they work. Velocity meters use the rate of flow to measure volume. Positive displacement meters are mechanical; the water physically moves, or displaces, the measuring mechanism. Compound meters combine two meters into one to measure large and small flows. Proportional meters measure only a portion of the total flow; total volume measurements are based on that sample. Open-channel meters are often built into a channel, such as a flume; they obstruct or restrict flow in order to measure it.

Velocity Meters

There are five types of velocity meters: propeller, turbine, ultrasonic, electromagnetic, and differential pressure. All convert velocity measurements into flow measurements.

Propeller. The measuring element in a propeller meter is a rotor, or propeller, facing upstream. The propeller is rotated by the moving water striking its angular blades. The propeller may be small (in diameter) in relation to the internal diameter of the pipe, especially in the larger sizes. The propeller has a slight lag in starting or stopping; therefore, this meter is intended primarily for main-line service where flow rates do not change abruptly.

Turbine. A turbine meter consists of a very light waterwheel operated by the water's current. The waterwheel carries on its axis a worm for actuating gearing and a totalizer. The rate of flow is computed from the rotations of the waterwheel during a given period. A more recent design, the magnetic flowmeter, requires electrical connections and is not a completely self-contained unit.

Ultrasonic. Ultrasonic, or acoustic, flowmeters operate on the principle that the propagation velocity of acoustic signals in liquids is changed when a component of the liquid's velocity parallels the direction of acoustic propagation. In practice, up to four acoustic paths are set up at an angle to the flow. Each path consists of two identical transducers facing each other at a precise distance. Typical path angles vary from 30° to 65°, depending on available space and accuracy requirements. The liquid's average velocity on each path is determined by measuring the acoustic travel time in each direction. The liquid velocity measured at each path is integrated across the flowing area to determine the total volume flow rate. Such meter systems have very critical installation requirements and, while good for computer input, are rather costly. Both ultrasonic and Doppler-flow measurement techniques require turbid rather than clear fluid to work effectively.

Electromagnetic. The electromagnetic flowmeter uses the same basic principle as the electric generator. When a conductor moves across a magnetic field, a voltage is induced in the conductor, and the magnitude of the voltage is directly proportional to the speed of the moving conductor. If the conductor is a section of conductive liquid flowing in a nonconductive pipe through a magnetic field, and electrodes are mounted in the pipe wall, the voltage induced across the electrodes should be proportional to the flow rate.

Differential pressure. Three types of meters are based on differential pressure.

Flow tubes/venturi. As fluid passes through the reduced area of the venturi throat its velocity increases, resulting in a pressure differential between the inlet and throat regions. The passage immediately following the throat gradually increases in flow area; consequently, the fluid's velocity decreases, causing pressure recovery. The differential pressure across the venturi's throat can be recorded directly or translated into actual flow units by using various types of differential pressure meters and capacity curves.

Venturi tubes are used whenever head loss must be minimized or where fluids contain enough materials in suspension to render other devices, such as orifice plates or flow nozzles, ineffective. Their widest application has traditionally been in low-pressure gas lines and large water mains.

Measurements accurate to within 0.5 percent can be made for nearly all pipe sizes. For larger-diameter pipes, piezometer rings and tap arrangements may be critical.

Venturi tubes maintain their accuracy over relatively long periods. In most applications the venturi is a self-cleaning device because its internal configuration allows smooth flow and minimizes erosion and clogging. Venturi tubes are mostly maintenance-free. They have no moving parts nor mechanical features or glass that can weaken or break. Testing the accuracy of the venturi is difficult, because alternative measuring devices of similar accuracy and size are often unavailable or difficult to install. When calibrating the venturi, it is imperative to check the venturi itself and not just the manometer apparatus. Be sure that the gauges are calibrated to agree with the actual confirmed flow. Do not adjust the manometer apparatus to agree with a hypothetical standard that probably does not occur.

Orifice plate. The orifice sensing unit consists of a round plate in which is bored, usually concentrically, a hole of predetermined size. The plate is inserted across a straight run of pipe. Pressure taps are provided, either at specified distances upstream and downstream from the orifice plate or within orifice flanges inserting the orifice plate in the line.

The relation between flow rates and differential pressures has been reduced to formulas by which an orifice coefficient can be readily computed for any size of orifice in any size of pipe.

Pitot rod. The pitot rod measures fluid velocities by its sensitivity to pressure differences associated with the pitot rod when it is inserted in the pipeline. Velocity measurements are made along a transect inside the pipe. Then the velocity measurements are multiplied by the proportional cross-sectional area to determine the fluid flow. The sum of all the proportional flows is the total flow within the pipe.

Positive Displacement Meters— Nutating Disc

The nutating disc meter consists of a movable disc mounted on a concentric sphere. The disc is contained in a working chamber with spherical side walls and top and bottom surfaces that extend conically inward. It is restricted from rotating about its own axis by a radial partition that extends across the entire height of the working chamber. The disc is slotted to fit over this partition. The water enters the side of the meter and strikes the disc, forcing it to rock (nutate) in a circular path without rotating about its own axis. A pin extending out from the inner sphere, perpendicular to the disc, traces a circular path as it is driven by the nutating motion. This pin drives the undergear that controls the meter's register.

Compound Meters

This meter measures over a wide range of flow, including very low flows. It is called *compound* because it has a large turbine meter on the main line and a small meter on a bypass line, with a valve to direct water to one or the other meter automatically. Sometimes both meters are in the same housing. Compound meters may have either single or double registers.

Proportional, or Fire-Line, Meters

The proportional meter measures a small but relatively constant percentage of water flowing through a line. A multiplying factor is built into the meter register so that it records the total quantity. This is accomplished by providing an expanded section of the line with a ring at the downstream end having an internal diameter equal to the nominal pipe diameter. The restriction ring creates a drop in line pressure just below it (compared with full line pressure upstream). This differential in pressure forces a portion of the water through the small displacement-type bypass meter.

Open-Channel Meters

There are two basic types of open-channel meters: weirs and flumes.

Weirs. A weir is an obstruction built across an open channel over which the water must flow. The water usually flows through an opening or notch on the weir plate. Sharp-crested weirs have three basic configurations: rectangular, V-notch, and Cipolletti.

Flumes. Flumes are specially designed, open-channel flow sections with a restriction that increases fluid velocity. They can be designed for installation in a circular pipe section.

Flumes have several advantages over weirs. The flow velocity through flumes is high; therefore, they tend to be self-cleaning systems that minimize deposition of sediment or solids. Flumes can operate with much smaller head losses than weirs, a decided advantage for irrigation.

A Parshall flume is used where it is important to maintain a low head loss or where the liquid contains a large amount of suspended solids.

RECOMMENDED USES OF METERS BY CLASSIFICATION

All meter installations should be reviewed to determine whether or not the proper meter has been chosen for each installation. Table B-1 lists uses of meters by classification. Table B-2 compares recommended uses for various types of meters. Refer to *Water Meters—Selection, Installation, Testing, and Maintenance** for additional information.

METER SIZING PARAMETERS

Once the type of meter has been selected, the meter should be properly sized for operation. This means that it must measure the water accurately through the full range of the customer's use. Water meters are designed to measure a maximum flow for short periods and a lower flow for long periods without sustaining damage or above-normal wear.

Revenue can be lost because a meter does not register accurately. Inaccurate registration can result from wear, improper installation, and improper size or type of meter.

When a meter operates outside its intended range, it cannot register all the flow, even though the meter itself may be calibrated and working accurately. For a compound meter with appreciable flow at the change-over point, the loss in revenue can be substantial.

A 24-hour or 7-day recorder attached to a water meter will provide maximum, average, and minimum rates of flow, which are essential in determining whether or not the meter is the proper size and type for the installation. Without comparing this information with the meter's capacity, selecting a proper size and type of meter becomes a matter of guesswork based on experience. For large-meter installations, a wrong guess can be costly.

Large water meters register flow inaccurately for several reasons, including the following:

- The wrong size or type of water meter was selected. In some cases, the meter was selected to match the size of the service. Although flow should not be restricted, considering the amount of revenue at stake, meter sizing should be based on the type of factual information provided by a recorder chart.
- A change in customers occurred at a specific location. The usage profile of the new customer often differs considerably from that of the original customer.
- A change in customer usage due to water conservation or changes in their production and/or products.

A meter should be of adequate size but not oversized. Sizing should be based on the maximum and minimum demand and average usage rate of the customer.

AWWA has established standards on flow capacities and maximum pressure losses for cold-water meters. In addition, some new meters cause lower pressure losses. The estimator can follow the AWWA standards or the head-loss flow curves provided by the manufacturer for newer models, as appropriate.

Tables B-3, B-4, and B-5 provide data from which approximate pressure loss can be determined. These tables are based on pressure-flow data taken from curves of

*AWWA Manual M6, 4th ed., AWWA, Denver, Colo. (1999).

Table B-1 Recommended uses of meters by classification

Type	Size *in.*	Applications
Positive Displacement (ANSI/AWWA C700*)	⅝	Residences, small apartments, small businesses. Demand flow rates: ¼–20 gpm. Maximum continuous demand to 10 gpm.
	¾	Large residences, small to medium apartments. Demand flow rates: ½–30 gpm. Maximum continuous demand to 15 gpm.
	1	Medium apartments, beauty parlors, barber shops, small motels, filling stations, small businesses, industrial processes. Demand flow rates: ¾–50 gpm. Maximum continuous demand to 25 gpm.
	1 ½	Medium motels, hotels, large apartments, small industry, small processing plants. Demand flow rates: 1½–100 gpm. Maximum continuous demand to 50 gpm.
	2	Larger hotels, motels, apartment complexes, industrial plants, processing plants. Demand flow rates: 2–160 gpm. Maximum continuous demand to 80 gpm.
Turbine— Class II (ANSI/AWWA C701†)	2	Medium to large hotels, motels, large apartment complexes, industrial plants, processing plants, irrigation. Demand flow rates: 4–160 gpm. Maximum continuous demand to 100 gpm.
	3	Large hotels, motels, industrial plants, processing plants, irrigation. Demand flow rates: 8–350 gpm. Maximum continuous demand to 240 gpm.
	4	Large industrial and processing plants, irrigation, refineries, petrochemicals, pump discharge. Demand flow rates: 15–630 gpm. Maximum continuous demand to 420 gpm.
	6	Large industrial manufacturing and processing plants, irrigation, pump discharge. Demand flow rates: 30–1,400 gpm. Maximum continuous demand to 920 gpm.
Turbine— Class I‡ (ANSI/AWWA C701†)	8	Industrial, manufacturing, processing, pump discharge. Demand flow rates: 140–1,600 gpm. Maximum continuous demand to 900 gpm.
	10	Industrial, manufacturing, processing, pump discharge. Demand flow rates: 225–2,500 gpm. Maximum continuous demand to 1,450 gpm.
	12	Industrial, manufacturing, processing, pump discharge. Demand flow rates: 400–4,000 gpm. Maximum continuous demand to 2,150 gpm.

*ANSI/AWWA C700-95, AWWA Standard for Cold-Water Meters—Displacement Type, Bronze Main Case.
†ANSI/AWWA C701-88, AWWA Standard for Cold-Water Meters—Turbine Type, for Customer Service.
‡Class I turbines smaller than 8 in. are not included because of the higher performance of class II models.

Table continued next page.

Table B-1 Recommended uses of meters by classification (continued)

Type	Size in.	Applications
Compound—New High-Velocity Styles (ANSI/AWWA C702*)		Medium motels, hotels, special customers having high and low demand: schools, public buildings, large apartment and condominium complexes, hospitals.
	2	Demand flow rates: ¼–160 gpm. Maximum continuous demand to 80 gpm.
	3	Demand flow rates: ½–320 gpm. Maximum continuous demand to 160 gpm.
	4	Demand flow rates: ¾–500 gpm. Maximum continuous demand to 250 gpm.
	6	Demand flow rates: 1½–1,000 gpm. Maximum continuous demand to 500 gpm.
	8	Demand flow rates: 2–1,600 gpm. Maximum continuous demand to 800 gpm.
	10	Demand flow rates: 4–2,300 gpm. Maximum continuous demand to 1,150 gpm.

*ANSI/AWWA C702-92, AWWA Standard for Cold-Water Meters—Compound Type.

Table B-2 AWWA C701* turbine meter standards for customer service

Meter Size in.	Safe Maximum Operating Capacity gpm	Maximum Rate for Continuous Duty gpm	Maximum Loss of Head at Safe Maximum Operating Capacity psi	Normal Test Flow Limits gpm
Class I—Vertical-Shaft and Low-Velocity Horizontal Type				
1½	100	50	15	12–80
2	160	80	15	16–120
3	350	175	15	24–250
4	600	300	15	40–400
6	1,250	625	15	80–1,000
8	1,800	900	15	140–1,600
10	2,900	1,450	15	225–2,500
12	4,300	2,150	15	400–4,000
Class II—In-Line (High-Velocity Type)				
2	160	100	7	4–160
3	350	240	7	8–350
4	630	420	7	15–630
6	1,400	920	7	30–1,400
8	2,400	1,600	7	50–2,400
10	3,800	2,500	7	75–3,800
12	5,000	3,300	7	120–5,000

*ANSI/AWWA C701-88, AWWA Standard for Cold-Water Meters—Turbine Type, for Customer Service.

Table B-3 Displacement-type meters meeting AWWA specifications: flow-pressure loss averages of 1973-model meters*

Size	Maximum Capacity AWWA Flow Criteria		Recommended Design Criteria—80% of Maximum Capacity		Recommended for Continuous Flow—30% of Maximum Capacity		Brands of Meters
in.	*gpm*	*psi*	*gpm*	*psi*	*gpm*	*psi*	Avgs.
3/8 × 3/4	20	10.4	16	6.1	6.0	1.0	6
3/4	30	10.6	24	6.9	9.0	1.05	6
1	50	9.3	40	6.3	16.5	1.0	6
1 1/2	100	11.3	80	8.6	30.0	0.9	6
2	160	10.4	128	6.5	38.0	0.5	6
3	300	13.1	240	8.3	90.0	1.1	3

**Sizing Water Service Lines and Meters*, AWWA Manual M22, AWWA, Denver, Colo. (1975).

Table B-4 Compound-type meters meeting AWWA specifications: flow-pressure loss averages of 1973-model meters*

Size	Maximum Capacity AWWA Flow Criteria		Recommended Design Criteria—80% of Maximum Capacity		Recommended for Continuous Flow—50% of Maximum Capacity		Brands of Meters
in.	*gpm*	*psi*	*gpm*	*psi*	*gpm*	*psi*	Avgs.
2	160	9.2	128	6.1	80	2.6	3
3	320	13.4	250	8.9	160	4.2	5
4	500	9.6	400	6.3	250	3.5	5
6	1,000	9.4	800	5.8	500	2.5	4
8	1,600	12.0	1,280	7.8	800	4.0	3

**Sizing Water Service Lines and Meters*, AWWA Manual M22, AWWA, Denver, Colo. (1975).

Table B-5 Turbine-type meters meeting AWWA specifications: flow-pressure loss averages of 1973-model meters*

Size	Maximum Capacity AWWA Flow Criteria		Recommended Design Criteria—80% of Maximum Capacity		Recommended for Continuous Flow—50% of Maximum Capacity		Brands of Meters
in.	*gpm*	*psi*	*gpm*	*psi*	*gpm*	*psi*	Avgs.
2	160	4.5	128	2.8	80	1.0	5
3	350	4.6	280	3.0	175	1.2	4
4	600	3.5	480	2.1	300	0.8	4
6	1,250	3.5	1,000	2.0	625	0.75	4

**Sizing Water Service Lines and Meters*, AWWA Manual M22, AWWA, Denver, Colo. (1975).

several types of 1973-model meters. The data were averaged to produce values that the estimator can use when rating curves are not available. These tables are not applicable to some earlier meters, and the estimator should use them only as approximations for late-model meters.

There are several important considerations in sizing a meter installation. The flow requirement, not the pressure loss through the meter, should determine the size of the meter to be installed. Oversizing a meter to lower the pressure loss can mean low flows will fail to register. It will also increase the cost of maintenance. Furthermore, if there is good reason to believe the customer plans to increase the demand for water, the estimator should provide for larger metering facilities later, adding a meter box and connections that will meet future needs.

METER INSTALLATION CONSIDERATIONS

Correct meter installation is very important. Because a meter is a mechanical device, it begins to wear and lose accuracy from the day it is installed. In time it will stop functioning. Provisions should be made during installation to field-test meters in line for accuracy.

When installing large meters, include the following:

- Isolation valves in both the inlet and outlet sides of the meter.
- Test plugs downstream of the meter between the meter and the downstream isolating valve.
- A meter bypass with isolating valve or a bypass meter to provide water to the customer while the meter is being serviced. The bypass connection should be placed upstream of the inlet isolating valve and downstream of the outlet isolating valve so that it forms a *U* around the main meter.

  If a meter is installed with a bypass meter, the bypass meter should remain open and a check valve must be installed downstream to force water at low flow through the bypass meter. If the meter is not installed in this manner, low flows will pass unregistered through the large meter as water follows the path of least resistance. The check valve should be either internally spring-loaded or externally controlled so that a differential pressure of about 5 to 10 psi is required to open the check valve.

  If meters are installed in parallel, one meter should act as a primary meter and the other as a secondary meter. To do this, place a check valve downstream of the secondary meter to ensure a pressure loss of at least 5 to 10 psi through the primary meter before the secondary meter begins operating. This provision is necessary to make certain both meters register low flows.
- To ensure proper meter functioning (by reducing turbulence), install the meters according to the manufacturers' specifications governing the number of straight pipe diameters upstream and downstream of the meter. Another method is to install straightening vanes.

When installing turbine meters, use a strainer on the inlet side of the turbine. The strainer has two functions: to keep foreign matter from damaging the turbine and to ensure that the water approaching the turbine is fairly equal in velocity across the diameter of the pipe and is not spiraling.

Appendix C

Meter Testing, Surveillance, and Calibration

A water meter, like any other mechanical device, is subject to wear and deterioration. Over a period of time, it loses its peak efficiency. How long water meters retain their overall accuracy depends on many factors, such as the quality of the water being measured, rates of flow, total quantity, chemical buildup, and abrasive materials carried by the water. The only way to determine if a specific meter is operating efficiently is to test it.

TESTING

Meters are tested for several reasons, including determining (1) the total amount of meter error within the distribution system, (2) if the cost of water service is equitably distributed among all customers, (3) the amount of revenue lost to the water utility, and (4) if a meter is capable of registering low flows. Unfortunately, displacement meters, the type most commonly used, may seriously underregister for long periods without completely ceasing to operate.

Elements of a Meter Test

The three basic elements needed to test overall meter efficiency are

- the number of different flow rates to be tested for the operating range of the meter
- the volume of water to be passed through the meter for each test at each designated flow rate
- accuracy limits that meters must meet at differing rates to be acceptable for use

Test Rates

Three rates of flow—minimum, intermediate, and maximum—are necessary to properly test displacement, compound, and propeller meters. For compound and fire-service meters, the changeover point must be established to determine overall operational efficiency and accuracy of registration. It is also important to establish at what point a meter begins to register low flow. Low flows below that point do not produce revenue.

The meter testing program should begin with low flow rates. If in the course of testing, the meter speed changes, make sure there was no use by the customer. High flows must be avoided before low- and medium-flow testing because the high flow could flush out sand and other materials. Such flushing would totally negate later low-flow tests.

The intermediate rate of flow should be at or near the high point of registration (about 10 percent of rated capacity) to ensure that the meter will not overregister at any rate of flow.

Tests for full-flow accuracy are not necessary at the safe maximum capacity rate shown in AWWA standards because meters are seldom operated at rated capacity. The maximum point of registration depends on meter design but is usually about 10 percent of rated meter capacity. At rates above that, the accuracy curve is fairly flat, and accuracy differs little over a wide range of flows. Maximum-rate test flows of about 75 percent of rated capacity are practical. They are especially advantageous in multiple testing of small meters.

Accuracy Limits

Accuracy limits are established to ensure that water meters record as accurately as possible and commercially feasible. Meters tend to register variably by 2 to 3 percent for the entire range of flows. This does not occur when the flow is so low that it fails to register. For example, a ⅝-in. meter in good condition will register within the following limits: at 95 percent or higher with a ¼-gpm flow, at a maximum of 101.5 percent with a 2-gpm flow (usually 10 percent of rated meter capacity), and at not less than 98.6 percent with a 20-gpm flow.

It is not considered economically feasible to repair older meters to bring them up to accuracy levels achievable by new meters at the minimum rate. For this reason, separate accuracy limits for the minimum flow test are given for new and repaired meters. Limits set for repaired meters represent good repair procedures. Repaired meters should register at least 90 percent on this test. A higher percentage is recommended for shop quality standards.

Meter Test Alternatives

Meters can be tested in place (field testing) or in the meter shop (bench testing).

Field testing. As the cost of meter removal and replacement increases, field testing becomes more attractive. Large meters can be tested with a pitot rod. The portable meter test unit is adequate for smaller meters. The portable test equipment compares a previously calibrated meter with the meter being tested. In field testing, both meters must be full of water and under positive pressure, and the control valve to regulate flow must always be on the discharge side of the calibrated meter.

Bench testing. A meter shop typically uses equipment that varies from a simple rate-of-flow indicator to a volumetric test bench that can test several meters at one time. Rising costs of meter removal and replacement are causing many utilities to phase out their meter shops and either replace their meters without testing or send them out to be tested.

For a detailed description of meter testing, refer to *Water Meters—Selection, Installation, Testing, and Maintenance.*

Meter Testing with a Pitot Rod

An insertion velocity probe is used to measure the velocity of fluid in a closed conduit under pressure. A specifically designed caliper measures the internal diameter of the conduit. From these measurements flow can be calculated to within about 2 percent accuracy, a greater rate of accuracy than can be attained with an averaging insertion velocity probe.

The insertion velocity probe is effective on any closed conduit that has a 1-in. corporation cock installed. An experienced operator can perform the test in about 2 h under good conditions.

Insertion velocity probes can be used to test the accuracy of large water meters, measure flow in a pipe, perform loss-of-head tests, perform district measurements, determine direction of flow, calculate 24-h flows, and indicate flow reversals.

LONG-TERM METER SURVEILLANCE

After a meter has been tested and calibrated, surveillance or review of consumption records should be initiated to ensure the meter will operate as long and as accurately as possible at the lowest possible cost. The method given here is suited for large meters and large revenue-producing customers.

One useful technique in meter surveillance is reviewing billing records at regular periods for each meter. Computerization helps immensely. Otherwise, sampling may be necessary. It is not necessary to review records every billing period; a quarterly or semiannual review is sufficient. Compare the records for at least a 12-month period to guard against seasonal changes in the customer's water-use pattern. Contact the customer and, if it is found that the customer's actual usage is not changing and the consumption records are declining, then the meter may be losing accuracy and should be scheduled for inspection and testing.

Flow rates vary from customer to customer, even when meters of the same size are in use. Rates of use also vary by season. It is useful to record each customer's typical maximum and minimum monthly usage to anticipate demand.

RECALIBRATING LARGE METERS

One of the major benefits of conducting a water audit is the potential increase in revenues from testing and repairing larger meters. One can estimate changes in revenue from large meters found to be operating inaccurately (see Table C-1).

The calculation for the corrected metered volume for each individual meter is the following:

$$CMV = \frac{UMV}{(0.01)MR} \qquad \text{(Eq C-1)}$$

Table C-1 Potential revenues from recalibrated large meters

(1) Meter ID Number	(2) Uncorrected Metered Volume *UMV*	(3) Charges Billed in 12 Months *$*	(4) Corrected Metered Volume *CMV*	(5) Charges for Corrected Volume for 12 Months *$*	(6) Increased Revenues (column 5 minus column 3) *$*
9793234	3,644	3,847	5,906	6,235	+ 2,388
20548059	5,141	5,751	8,199	9,172	+ 3,421
X123456	6,668	6,911	19,328	20,031	+13,120
16222519	5,738	5,860	26,939	27,512	+21,652

Where:

CMV = corrected meter volume
UMV = volume of water registered on the meter (from billing records)
MR = mean registration, or accuracy, in percent (determined by meter tests)

NOTE: Any unit of measure may be used for UMV. The CMV measurement will be the same. See example.

Example:

$$UMV = 10{,}000 \text{ ft}^3 \text{ (from billing records)}$$

$$MR = 75\% \text{ (registration accuracy determined by meter tests)}$$

$$CMV = \frac{10{,}000}{(0.01)(75)}$$

$$= 13{,}333 \text{ ft}^3$$

Table C-1 shows revenues that can be recovered by recalibrating and repairing large meters for customers who have large volumes of water use. Frequently, these revenues can pay for substantial portions of the water audit as well as the meter test and repair.

Appendix D

Calculating Evaporative Water Loss

Evaporative water losses from large bodies of water include water lost directly to the atmosphere from open-air, standing bodies of water. These losses are particularly significant during hot weather at large unmetered pools and reservoirs. For example, ponds and reservoirs in California annually lose an average of 50 to 80 in. of water through evaporation. It is often difficult to estimate losses from small pools and fountains.

During the summer, evaporation losses may be slightly under 0.33 in./day from each acre of pond or reservoir surface, or about 9,000 gal/day/acre. Exposure to winds and high air temperatures are the principal contributing factors. Wind disrupts the saturated boundary layer over the water surface, and air increases its capacity to hold moisture as its temperature rises.

Calculations of evaporative losses from ponds, lakes, and reservoirs are complex and require detailed local environmental data. Additional data for calculating evaporative water loss may be found in the two references listed below, or from state and provincial agricultural services.

REFERENCES

Bennett, R.E., and M.S. Hazinski. 1993. *Water-Efficient Landscape Guidelines*. Denver, Colo.: American Water Works Association.

Toro Company. 1966. *Rainfall-Evapotranspiration Data, United States and Canada*. Minneapolis, Minn.: Toro Company.

Appendix E

Leak Detection Equipment

For information on water audits and leak detection suppliers and consultants, refer to the *AWWA Sourcebook*, the official resource guide to water industry products and services. For a free copy of this guide, call 1-800-926-7337.

Appendix F

US Customary to Metric Conversions

To Convert From US Customary	To Metric	Multiply by
acres	hectares	0.4047
acres	square metres	4,047
acre-feet	cubic metres	1,233
feet	metres	0.305
cubic feet	cubic meters	0.02832
gallons	cubic metres	0.003785
gallons	litres	3.785
gallons per minute	litres per second	0.063
gallons per day	litres per day	3.785
inches	millimetres	25.4
square inches	square millimetres	645.2
pounds	kilograms	0.4536
pounds per square inch	kilopascals	6.895
miles	kilometres	1.609

Index

NOTE: An *f.* following a page number refers to a figure; a *t.* refers to a table.

AWWA List of Manuals

M1, *Water Rates*, Fourth Edition, 1991, #30001PA

M2, *Automation and Instrumentation*, Second Edition, 1983, #30002PA

M3, *Safety Practices for Water Utilities*, Fifth Edition, 1990, #30003PA

M4, *Water Fluoridation Principles and Practices*, Fourth Edition, 1995, #30004PA

M5, *Water Utility Management Practices*, First Edition, 1980, #30005PA

M6, *Water Meters—Selection, Installation, Testing, and Maintenance*, Second Edition, 1999, #30006PA

M7, *Problem Organisms in Water: Identification and Treatment*, Second Edition, 1995, #30007PA

M9, *Concrete Pressure Pipe*, Second Edition, 1995, #30009PA

M11, *Steel Pipe—A Guide for Design and Installation*, Fourth Edition, 1989, #30011PA

M12, *Simplified Procedures for Water Examination*, Second Edition, 1997, #30012PA

M14, *Recommended Practice for Backflow Prevention and Cross-Connection Control*, Second Edition, 1990, #30014PA

M17, *Installation, Field Testing, and Maintenance of Fire Hydrants*, Third Edition, 1989, #30017PA

M19, *Emergency Planning for Water Utility Management*, Third Edition, 1994, #30019PA

M20, *Water Chlorination Principles and Practices*, First Edition, 1973, #30020PA

M21, *Groundwater*, Second Edition, 1989, #30021PA

M22, *Sizing Water Service Lines and Meters*, First Edition, 1975, #30022PA

M23, *PVC Pipe—Design and Installation*, First Edition, 1980, #30023PA

M24, *Dual Water Systems*, Second Edition, 1994, #30024PA

M25, *Flexible-Membrane Covers and Linings for Potable-Water Reservoirs*, Second Edition, 1996, #30025PA

M26, *Water Rates and Related Charges*, Second Edition, 1996, #30026PA

M27, *External Corrosion—Introduction to Chemistry and Control*, First Edition, 1987, #30027PA

M28, *Cleaning and Lining Water Mains*, First Edition, 1987, #30028PA

M29, *Water Utility Capital Financing*, Second Edition, 1998, #30029PA

M30, *Precoat Filtration*, Second Edition, 1995, #30030PA

M31, *Distribution System Requirements for Fire Protection*, Second Edition, 1992, #30031PA

M32, *Distribution Network Analysis for Water Utilities*, First Edition, 1989, #30032PA

M33, *Flowmeters in Water Supply*, Second Edition, 1997, #30033PA

M34, *Alternative Rates*, Second Edition, 1999, #30034PA

M35, *Revenue Requirements*, First Edition, 1990, #30035PA

M36, *Water Audits and Leak Detection*, Second Edition, 1999, #30036PA

M37, *Operational Control of Coagulation and Filtration Processes*, First Edition, 1992, #30037PA

M38, *Electrodialysis and Electrodialysis Reversal*, First Edition, 1995, #30038PA

M41, *Ductile-Iron Pipe and Fittings*, First Edition, 1996, #30041PA

M42, *Steel Water-Storage Tanks,* First Edition, 1998, 30042PA

M44, *Valve Installation and Maintenance*, First Edition, 1996, #30044PA

M45, *Fiberglass Pipe Design*, First Edition, 1996, #30045PA

M47, *Construction Contract Administration*, First Edition, 1996, #30047PA

To order any of these manuals or other AWWA publications, call the Bookstore toll-free at 1-(800)-926-7337.